The Smallholder's Manual

Katie Thear

The Crowood Press

First published in 2002 by
The Crowood Press Ltd
Ramsbury, Marlborough
Wiltshire SN8 2HR

www.crowood.com

This impression 2006

© Katie Thear 2002

British Library Cataloguing-in-Publication Data
A catalogue record for this book is available from the British Library.

ISBN 1 86 126 555 7

EAN 978 1 86126 555 5

Acknowledgements
I have been involved with smallholdings for many years, not only as a practising
smallholder, but also as an author and journalist. I have written about our activities
in a number of books, including *Part-Time Farming* (Ward Lock), *The Family
Smallholding* (B. T. Batsford) and *A Kind of Living* (Hamish Hamilton/Channel 4
TV). I have also written regular articles for *Country Smallholding* magazine, and
visited smallholdings all over the world. A distillation of these sources and
experiences provides the basis for this present work. A framework, however, is
incomplete without up-to-date adornments. Consequently, the book includes all
the latest ideas and techniques, as well as the current requirements for living and
working on a modern smallholding.

Many individuals and organizations have provided information, suggestions
and advice, and I am grateful for their unstinting co-operation. Relevant companies
and organizations are listed in the reference section at the end of the book.

Finally, my grateful thanks go to my husband David and to our children
Matthew, Helen and Gwilym for their love and active help through the years.

Illustration Credits

The photographs are taken by the author, andd courtesy of the following:
Clearview Stores (p.18), Aga (p.20), Conder (p.29), Agria (p.34),
Stihl (p.36), Elsoms (p.63), Thompson and Morgan (p.68), Pan (p.71),
B & Q (p.79), Bayliss (p.79), Clovis Lande (p.80), Otley College (p.104),
DEFRA (p.110), Associated Poultry (p.118), Hubbard-ISA (p.123),
Sorex (p.132), H.D. Sharman (p.145), Early Bird Products (pp.155 and 159),
John Adlard (p.159), Hengrave Feeders (p.162), Sturdy Sties Ltd (p.176),
Cotswold Pig Development-Co. (pp.177 and 179), Chris Sowe (p.193),
Boer Goat Society (p.194), The British Wool Marketing Board (p.217).

Some of the line drawings are reproduced from the author's previous works;
they were drawn from her original sketches by Jane E. Smith, G.L. Galsworthy,
Frederick St. Ward, Graham Turner and Nils Solberg.

Typeset by Textype, Cambridge

Printed and bound in Spain by Mateu Cromo, Madrid

Contents

Preface

Make good use of modern developments and technology whilst retaining the best of traditional lore and practice.

The aim of this book is to give practical information about the realities of living and working on a modern smallholding. The term 'smallholding' refers to a home with a small area of land where the activities are significantly associated with agriculture. The information is applicable to a wide range of differing situations, and I have set our experiences against the broader spectrum of small farming in general.

The problem of earning a living is also considered, for it is unlikely that a small acreage will provide an adequate income, unless there is a high degree of specialization. If run efficiently, it will certainly reduce living costs by providing a large measure of home-produced food, and it may also be a useful source of supplementary income. In our case, starting a small publishing business based at our premises solved the problem. The birth and development of the magazine *Country Smallholding* closely followed the development of our smallholding.

From the outset, we avoided any degree of 'factory' farming and kept our livestock in humane conditions. Incidences of BSE and virulent outbreaks such as foot and mouth disease have demonstrated all too well the necessity of smaller scale, sustainable systems. This does not mean turning back the clock. All our cultivations have been organic, but our practice has also been to make the best use of modern developments and technology whilst retaining the best of traditional lore and practice.

The Smallholder's Manual is essentially a practical handbook for the twenty-first century, and is suitable for anyone considering buying a small farm, or making sustainable use of an existing one.

Katie Thear, Newport, 2002

Introduction

One deep conviction was that we would never again rely on someone else for a living.

Our involvement with a smallholding came as a result of various factors. I was born on a Welsh smallholding, which had been in my family for generations. My mother ran the smallholding in the traditional way, with two cows, a few sheep and pigs, poultry and a large kitchen garden. We did not live solely from the smallholding, for my father, like many of the men in that part of North Wales, was a merchant seaman.

After schooling and a period of higher education, I spent several years teaching biology and rural studies. Then I met and married my husband David. There followed a period of increasing material plenty. David had rapidly worked his way up to being a director for a firm of magazine publishers. Matthew was our first-born, and Helen followed two years later. We led extremely comfortable and ordered lives. Then the blow fell. David was suddenly made redundant, and we woke to find ourselves with large financial commitments and no money, realizing for the first time how vulnerable we were to the tides of fortune. One quite deep conviction was that whatever we did, we would start our own business and never again rely on someone else for a living.

We started a small publishing agency and within two years we had not only come through the difficult period, but were enjoying a similar standard of living to that which we had before. Our attitudes, however, were quite different. We had realized, during a time when we were trying to produce more of our own food, how little practical information there was in print, which told people how to look after half-a-dozen hens or a chemical-free kitchen garden. We wondered whether we could do something about it.

We began to research the whole field of small-scale farming. At that time, I was also working for the British magazine *The Ecologist*, well known and respected for its publication *A Blueprint for Survival* that predicted so accurately the global, environmental damage that we are familiar with today. This was an extremely valuable experience, for it brought me into contact with eminent environmentalists such as the late Dr Schumacher, the author of *Small is Beautiful: A Study of Economics as if People Mattered.* The destructive effects of intensive farming methods on the countryside really came home to us, perhaps for the first time. We made the decision that we would publish our own magazine for smallholders, and that it would concentrate on natural farming and growing methods, and on keeping livestock humanely. In association with this, we decided to buy and run our own experimental smallholding. Then I realized I was pregnant. We were to have a new home and give birth to a baby and a magazine in the same year.

We found Broad Leys in a small village in north Essex, and decided that it suited our needs. There was a house, a separate cottage called The Bothy, destined to become the office, a few outbuildings and 2 acres of land. The local schools had a good reputation, there was a quick train service to London, and Cambridge, with its excellent cultural facilities, was only a few miles in the other direction.

Soon the first issue of the magazine was complete, but there was to be no respite, as I was already working on the second issue before the first issue was printed. Meanwhile I made arrangements for my baby to be born at home. Gwilym was a fine, healthy boy. While I was in labour I was correcting page proofs for the second issue of the magazine; as each contraction came, I temporarily gave up reading.

The second baby, the first issue of *Practical Self Sufficiency* was born two weeks later. A flood of letters poured in, full of encouragement and positive suggestions, as well as articles on every aspect of smallholding practice. There were a few of the other kind, of course, including a tirade from someone who apparently did not approve of 'women farmers'. One of the nicest letters was from an old lady who had first kept chickens before World War I; her knowledge and experience were extensive, and typical of the high standard of material that came in from ordinary country people all over Britain.

We had to reprint the first issue of the magazine

three times in order to meet the demand for copies. Since then, it has continued to develop, first into a bi-monthly publication called *Home Farm* and then into its present form, a national monthly magazine called *Country Smallholding*.

Part 1

The Rural Property

1 Buying a Smallholding

Is rural life really so idyllic?

The first essential is to concentrate on having an adequate income. Where this comes from will vary: it need not necessarily be from a full-time job. Many people now living in rural areas are finding that a part-time job, in conjunction with their smallholding activities, produces a comfortable and satisfying lifestyle. The job may be working for someone else or operating a small home business. A business can operate from a smallholding, without necessarily having anything to do with agriculture.

The second priority is to ensure that the house itself is adequate. Repairs, renovations and extensions to provide comfortable accommodation take precedence over a gardening programme or the acquisition of animals. Outbuildings and fences are third in order of priority, and only when these have been repaired and appropriately adapted, is it prudent to think about keeping livestock.

Location

Is the location suitable for everyone? While a country upbringing is generally good for children, it may also be a lonely one if the house is in an isolated area. How far is the nearest town, and what is the public transportation like? In some villages the public transport may be non-existent, and the availability of shops, schools and services may also leave a lot to be desired. Where local sales of farm produce are envisaged, easy access to ready markets is essential if a lot of time is not to be spent delivering the produce.

Cultural and linguistic factors should also be borne in mind. Some areas of Wales, for example, are completely Welsh speaking and may not welcome incomers who make no attempt to learn the language or respect the culture. It is also appropriate to mention the myth of 'town versus country', where the inhabitants of each group are supposedly incapable of understanding each other. The fable of the town and the country mouse portrays the characters as being out of their depth and unable to cope when taken out of their usual environments; thus the country frightens the town mouse, while the country mouse cannot wait to get back home from the town. It has always seemed to me that the intelligent mouse is the one who experiences and respects the town *and* the country, and therefore benefits from both worlds. There is nothing about the practices of town or country that cannot be absorbed quite quickly by those willing to adapt.

It is always a good idea to go and see a property

on a cold, wet, dismal day. Sunshine and cottage garden flowers can all too often disguise the realities of a north-facing site at the mercy of the prevailing winds, or the major structural defects of a building.

Climate and land topography play important roles in its suitability for crops and livestock. Prevailing winds can have a major effect on certain crops or young animals. Hill farms often have rough grazing and are suitable only for mountain sheep. Steep slopes can be inaccessible or dangerous for tractors. North-facing slopes can be a problem because of their exposed nature and lack of sun, while water-logged land should be avoided.

It is also worth checking on the shape and size of fields to establish their accessibility and workability. A good map will help to establish geographical features, as well as the existence of any public foot-paths or rights of way. A friend of mine had a footpath that ran past his kitchen window. An other-wise delightful old man from the village would frequently stop there and pee in the hedge, but my friend never had the heart to protest, although he did hint that the compost heap might benefit!

It is easy to establish what the pattern of weather has been in a particular area. The meteorological office and civil airports keep records and will provide the information on request. It is appropriate to remember, however, that global warming is bringing in quite rapid meteorological changes, and these may have a bearing on future plans.

Quality of the Land

The quality of land varies from one area to another. It may be heavy clay, quick-draining sand, thin chalky soil, acid peat or, if you are lucky, a friable medium-loam soil. The better soil will grow better crops, and this fact is usually reflected in the comparative land values, with the best agricultural land fetching the highest prices. The agriculture ministry of most countries publishes land classi-fication maps that indicate the type and quality of soils in the different regions. In the USA, there are 18,000 different types of soil, but the local extension agent will provide information specific to an area. In Britain, agricultural land is classified into five grades (*see* panel).

Soil testing will establish the type and nature of the soil, as will an examination of the prevalent weeds: for instance, rushes, *Juncus* species, denote waterlogging, while a profusion of sheep's sorrel, *Rumex acetosella*, is a sure sign of an acid soil. (*See* 'Kitchen Garden', p. 52).

It is a good policy to observe what neighbouring

Classification of Agricultural Land	
Grade 1	The best quality: suitable for most horti-cultural and agricultural uses.
Grade 2	Not quite as good. Problems may be encountered with some root crops, such as carrots, but generally suitable for most purposes.
Grade 3	Good quality crops, such as root crops, may be difficult to grow, but good for grazing and, depending on climate, suit-able for cereals.
Grade 4	Suitable mainly for grazing.
Grade 5	Rough grazing only.

farmers are doing with their land, because if they are dependent upon it for their livelihoods, you may be certain that they are not growing crops totally unsuited to the environment.

How Much Land?

How much land is required for a smallholding? This is rather like the question of how long is a piece of string, since it depends on what the owner wishes to do with the land, and how much money is available. A great deal can be achieved on a relatively small area of land, and it is generally true that the less land one has, the more carefully managed and productive it is in relation to its size. There is far more wastage involved where a large acreage is concerned.

A large garden can have a productive kitchen garden, greenhouse and hives of bees, as well as rabbits, chickens or ducks. As part-time activities, these can provide a certain amount of supplementary income, as well as providing for one's own situation.

A small field adjoining a garden would enable a couple of dairy goats to be kept, or a few pigs to be reared. A hectare (just under 2.5 acres) with good pasture would allow a small breeding flock of sheep to be kept, while 2 hectares (5 acres) and over would cater for larger livestock such as a cow and calf, llamas or alpacas, or an organic enterprise. Four hectares (around 10 acres) makes possible the growing of hay, with perhaps space given to a cereal crop or a range of forage crops for the livestock. A part of this might also be used for an orchard, woodlot or wood coppicing area.

It is easy to forget the pasture requirements of grazing animals. It is not enough that there is a

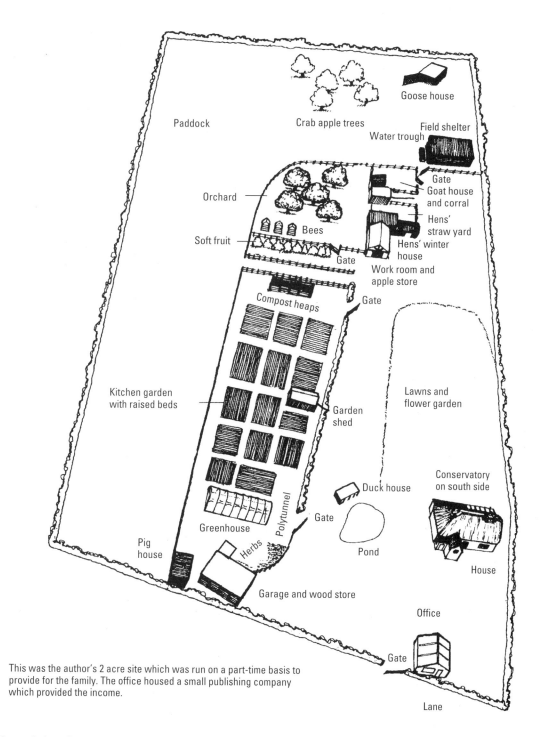

Goose house

Paddock

Crab apple trees

Field shelter

Water trough

Gate

Goat house and corral

Orchard

Bees

Hens' straw yard

Soft fruit

Gate

Hens' winter house

Work room and apple store

Gate

Compost heaps

Lawns and flower garden

Kitchen garden with raised beds

Garden shed

Conservatory on south side

Duck house

Pig house

Greenhouse

Gate

Herbs

Polytunnel

Pond

House

Garage and wood store

Office

Gate

This was the author's 2 acre site which was run on a part-time basis to provide for the family. The office housed a small publishing company which provided the income.

Lane

Ground plan of Broad Leys, the author's smallholding.

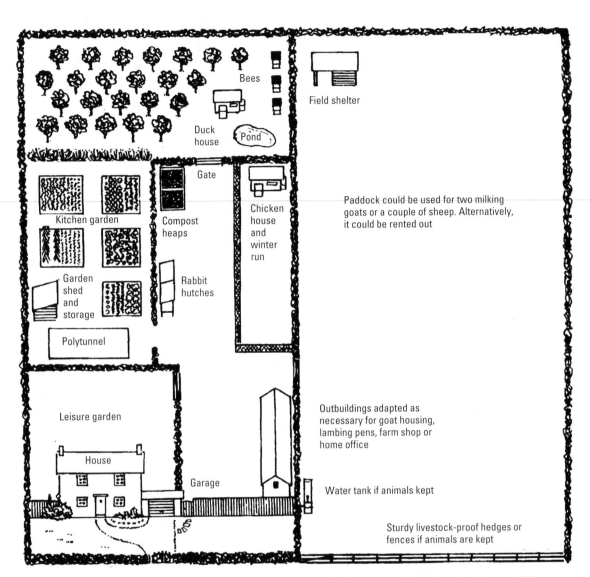

Bees

Field shelter

Duck house

Pond

Gate

Kitchen garden

Compost heaps

Chicken house and winter run

Paddock could be used for two milking goats or a couple of sheep. Alternatively, it could be rented out

Garden shed and storage

Rabbit hutches

Polytunnel

Leisure garden

Outbuildings adapted as necessary for goat housing, lambing pens, farm shop or home office

House

Garage

Water tank if animals kept

Sturdy livestock-proof hedges or fences if animals are kept

This part-time smallholding could have a non-agricultural enterprise operating from a home office. Alternatively the outbuildings may house a farm shop or animal housing. The orchard, kitchen garden and hens supply the home but surpluses could be sold at the farm gate. The orchard provides extra grazing for chickens in summer.

Rural property with large garden and small paddock.

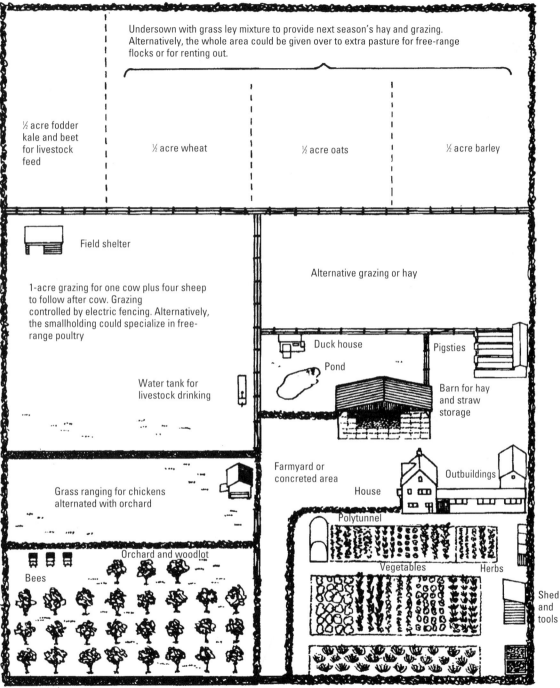

Undersown with grass ley mixture to provide next season's hay and grazing. Alternatively, the whole area could be given over to extra pasture for free-range flocks or for renting out.

½ acre fodder kale and beet for livestock feed

½ acre wheat

½ acre oats

½ acre barley

Field shelter

1-acre grazing for one cow plus four sheep to follow after cow. Grazing controlled by electric fencing. Alternatively, the smallholding could specialize in free-range poultry

Water tank for livestock drinking

Alternative grazing or hay

Duck house

Pigsties

Pond

Barn for hay and straw storage

Farmyard or concreted area

Outbuildings

House

Grass ranging for chickens alternated with orchard

Polytunnel

Vegetables

Herbs

Shed and tools

Orchard and woodlot

Bees

Entrance wide enough for vehicle and trailer access

The possibilities of a 5 acre smallholding.

Questions to Ask before Buying

Is the house suitable for your needs?
Can it be easily extended?

Is there mains or private sewerage?
What is its condition?

Is it a listed building?

Does it have all the necessary services?

Is there an 'agricultural use only' condition?
Can a non-agricultural business be started?

Are there any other restrictive covenants?

Is there a mains or private water supply?
What is its condition?

Can the water supply be extended?

Is there a private or a public electricity
supply? What is its condition?

Is three-phase electricity available for
heavy machinery?

How high is the farm?

Is it north- or south-facing?

Is it on steep ground?

Is it sheltered or exposed?

How far is the nearest town and shops?

How far are the nearest suitable schools?

How far is the nearest doctor and hospital?

How far is the nearest vet?

Is effective public transport available?

Is there easy car access to a main road?

What is the state of the outbuildings?

Is there evidence of vermin?

What grade agricultural land is it?

What is the pH value of the soil?

What are the prevalent weeds?

Are there many rushes and reeds?

What is the drainage like?

What is the condition of the hedges and ditches?

Are there any shared access rights?

Are there any public footpaths or rights of way across
the land?

What is the condition of fences and gates?

What trees are present on the site?

What is the average rainfall?

Is there any game on the land?

Are the fields of a convenient size and shape?

Are there any riparian (fishing) rights if there is a
stream running through the property?

Is there a local farmers' market?

Is there a local source of casual or part-time labour?

Is the farm in a tourist area?

Is it in a conservation area?

What are the neighbours like?

What leisure facilities and job prospects are available
for children as they get older?

What cultural facilities are available?

Is it really suitable? Be honest!

Broad Leys, the author's smallholding. The building on the right was the office.

certain area of grassland: there needs to be enough to allow for rotational grazing, so that as one area is used up, fresh ground is made available while the first is left to rest and recover. If hay is to be made, this represents even more pasture. Some smallholders find it more economic to use or rent out their grassland for grazing, while buying in hay for their own use.

The time and energy factors should not be overlooked. If priority is given to earning a living, there may be little of either to spare for looking after a collection of animals. The diagram on page 9 shows our 1.75 hectare (2 acre) site, and how it was used. The two goats, chickens, ducks, geese and bees were permanent residents, while pigs and a couple of sheep were reared for limited periods, but at different times. Our paddock was used for rotational grazing for the goats and geese, while the hens free-ranged in the orchard. The ducks ranged on the lawn area, as did the geese from time to time. We aimed to cover all the feeding and other costs and provide for ourselves, but not to make a profit, as our income came from another source.

The diagram on page 10 shows a rural property with a large garden and a small paddock developed purely for family use. The diagram on page 11 shows the theoretical possibilities of a larger smallholding run on a part-time basis, with an income coming from another source. These are obviously guidelines only, for a great deal depends upon the land, the inclinations of the owner, and the amount of time and capital available.

Particular Needs

Part-time smallholders are, by definition, part-time something else, and this is usually the half that generates an income. A home office is therefore often an essential pre-requisite on a modern smallholding; it may be in the home or in a converted outbuilding. Whether an enterprise is land-based or not, it is important to check that there are no restrictive covenants on the house or land that might curtail particular activities, such as keeping poultry. Conversely, if there is an 'agriculture only' designation, it may prove difficult to start a non-agricultural business, or to go in for a particular kind of building. Some countries have zoning, licensing and registration requirements; these vary in different states, but federal, state, county and city agencies will provide the necessary information. Legal searches will normally unearth any covenants or other conditions. It makes sense to establish whether there is likely to be a problem before the property is purchased; the panel opposite lists the main questions to ask.

2 The House

A rundown property has the potential for incorporating environmentally friendly features.

Each house will, of course, differ in its priorities, depending upon its condition and the needs of the occupants. Our particular experiences may be of interest to those thinking of embarking on a similar venture, but it is also appropriate to look at more general considerations. I am grateful to David Hills, an energy consultant, for his help and advice with this section of the book.

Condition

Few country properties that come up for sale are in an ideal condition in the buyer's eyes, and if they are, they are probably too expensive to be considered. However, a rundown property has considerable potential, not only to turn it into a comfortable family home, but also to incorporate environmentally friendly features. The immediate priorities are checking the condition of the basic structure – foundations, timbers, walls, windows and roof. Is there a damp-proof course? Are there any missing roof tiles? What is the condition of metal flashings around chimneys and dormer windows? Are the guttering and down-pipes in good condition? It is always worth checking the inside pipes. If possible, turn on all the taps and flush the lavatories.

Establish that the property has the necessary services – water, electricity, gas, sewage disposal and telephone. It is also important to establish the presence or otherwise of toxic materials such as asbestos or lead piping. A professional survey should reveal these so that a specialist contractor can remove them. In some areas, especially where granite is predominant, radon gas from the ground may be a problem. Environmental protection agencies will detect this, and advise on how buildings can be adapted to reduce its effect.

Most houses are built of bricks, blocks or stone. Our house was a timber-framed building, in-filled with lath and plaster, and topped with straw thatch. It appeared to have grown in stages over the years,

from what was originally a sixteenth-century yeoman's house. The roof leaked, the floorboards creaked, and the wiring gave every impression of having been installed by Faraday. We saw the immediate needs as being electrical repairs and renovations, the provision of extra bedroom space, thatched roof repairs, kitchen modernization, and general decoration.

Where renovations are concerned, a popular way of financing a project is to extend the mortgage. There are specialist building societies that lend money on the renovation and construction of houses that incorporate energy-efficient features, as well as on organic smallholdings and farms. It is necessary to work out how much repairs and renovations are likely to cost, and how long it will all take. In remote areas there may be a shortage of skilled labour, or at best, a long waiting list. If the house is a listed building, or is in a conservation area, there may be restrictions on the type and nature of any repairs, restorations or extensions. The local planning department will advise, as well as specifying the type of materials that may be used. For some projects, such as roofing or insulating, there may be a grant available.

It is likely that some or all of the timbers in a rundown property will need to be treated. The immediate task is to improve ventilation, remove badly affected timbers and, once the cause has been identified, treat with an appropriate fungicide or insecticide. It is a job for the professional. If there is an existing bat roost in the roofing space, the bats should not be disturbed, for they are a protected species in Britain and removing them is against the law. They are harmless, interesting creatures, and proofing treatment products that are not harmful to them are available. Prevention is a different matter, and screening can be put across the vents to stop any newcomers from entering.

If new timbers are required, avoid tropical hardwoods, unless they are known to be from a sustainable source.

If planning a complete renovation, a loft conversion may be well worth considering as the

Timber Problems	
Problem	**Comment**
Dry rot	Found in badly ventilated and damp conditions. Pungent smell, with dark cracks appearing in the timber. Grey patches with yellow or light purple tinges. If damage is extensive, professional help is needed to cut out decayed timbers and treat the remainder.
Wet rot	Apparent in wetter situations. Timber develops horizontal cracks with white or yellow markings. Easier to treat than dry rot. Effective ventilation usually sufficient, once affected areas have been removed and treated.
Wood-boring weevils	Feed on damp and decaying wood, leaving 1mm wide holes.
Common furniture beetle	Attacks all wood, not just furniture, leaving 1.5mm wide holes.
Death-watch beetle	Generally in old buildings. Tell-tale flight holes 3mm wide.
Powder beetle	1.5mm exit holes with a trickle of powdery dust.

environmentally friendly way to increase the living space. This is probably best left to a specialist contractor. If the plans to extend are rather more adventurous, then consider consulting an architect who specializes in ecological designs.

Recycled building materials are available from dismantled buildings at salvage yards around the country. Attractive as it may seem to use original timbers, bricks and roofing materials, they can be expensive. Modern bricks and tiles are available at building suppliers, which are replicas of the originals, at a fraction of the price. In some situations, though, where it is necessary to match a specific item, such as a chimney pot or uniquely patterned brick, the salvage yard is the best bet.

Some renovations may take longer than anticipated. A few weeks after we moved in, the suspicious-looking brown patch on the kitchen ceiling proved to be what we suspected: a place where water came in whenever it rained. Fortunately it was above the sink, but it was a nuisance. We immediately contacted the thatcher, thinking that we could put up with the dripping water for a week or two, or possibly even for a month or so if he was particularly busy. How naive we were about the mysteries of thatching supply and demand. The thatcher told us on the telephone that he did not need to come and examine the thatch for he knew the precise condition of it, just as he knew every other thatched building in the district. He explained that he would come and see us in two years' time. We had to reconcile ourselves to plopping water for some time to come.

The work involved two new hips or side slopes, a replacement layer for the south side, and a new ridge. The north side was virtually undamaged. Thatched roofs deteriorate much more quickly on the south-facing side because of the effects of the sun. Most roofs only require the top portions to be removed (the depth depending on the degree of damage), having new material inserted and then the whole thing 'combed' and netted. The wire netting is important because it ensures that birds do not get into the straw to nest. Our old roof had an extensive population of twittering fledglings, particularly in the section of roof above Matthew's window. In fact, once the roof was rethatched, he said that he missed the familiar sound.

Insulation

Renovating an old house provides a unique opportunity to create an energy-saving home. Half of the energy expended in the home goes on heating, although insulating and draught-proofing reduces the amount needed. Choosing a heating system that consumes as little energy as possible and that minimizes emissions also helps to reduce ozone depletion and global warming. It is possible to have a professional assessment of how energy efficient a home is. In Britain, for example, there is a national home energy-rating scheme (NHR) run by the National Energy Foundation. This gives a rating for the house on a scale of 1–10, where the higher the number, the greater the efficiency. The document provided after assessment is a useful addition to the property description should the house subsequently be sold.

Insulating the outside keeps the fabric of the house dry, warm and protected from the weather. If this is not practical, as for example where you need to leave exterior brickwork exposed, then walls can be insulated inside by attaching insulating board to interior walls or by attaching an inner wall onto battens with insulating material in between.

It is worth investing money in loft insulation, as this can save as much as 20 per cent on heating bills. There are three main types of loft insulation: mineral wool rolls, granular filler, and cellulose fibre which is made from fire-proofed recycled paper that has the added bonus of having consumed a minimal amount of energy in its production. Any insulation material should leave eaves entry points in the roof uncovered to ensure adequate ventilation and avoid condensation and damp.

Double-glazing can prevent as much as 20 per cent of the heat lost in an uninsulated home. Plastic sheeting is the cheapest option, but this needs to be replaced annually and is rather awkward. Secondary double-glazing involves the fixing of a second pane of glass (or plastic) to the window frame. There are various DIY kits available, as well as factory made, hermetically sealed, double-glazed units – these are the most expensive. It may be that the actual windows need replacing, in which case it may be worth choosing triple-glazed window units. A further energy-saving option is the use of Low-E (low-emissivity) glass that reflects heat back into the building.

Until we could afford permanent structures, we made our own double-glazing units for the north-facing windows. These were simple wooden frames fitted with 'trans-superglaze' polythene, a clear material, and held in place by thumb catches. They were effective in cutting out draughts and, in fact, worked well beyond our expectations. As they were light, and easy to remove and store, it was a simple matter to remove them in late spring when the weather was warm, and to re-install them in the autumn. We installed similar ones in the office. When we were able to afford more sophisticated double glazing at a later stage, the frames were put to use as large cloches in the kitchen garden.

The outer doors were provided with porches that gave excellent protection and insulation so that the house became draught free and cosy. The roof needed no insulation, for thatch is like a giant tea cosy, keeping warmth in, in winter, and providing cool conditions in summer.

In areas where winters are particularly cold, or there are severe winds, it is worth installing window shutters for added insulation and protection.

Heating Systems

A well insulated house will have a reduced need for heating, and a partial central heating system may be perfectly adequate. In our case, an open fireplace in

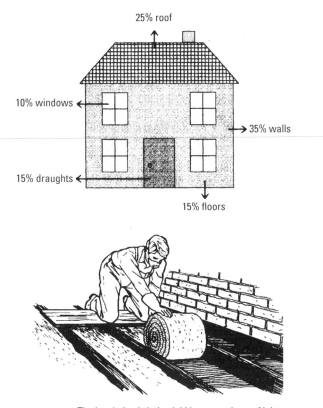

25% roof

10% windows

35% walls

15% draughts

15% floors

The insulation is being laid between the roof joists

Boarding over a batten framework filled with insulating material

How heat is lost in the home.

A warm and comforting multi-fuel stove.

the living room, an oil burner in the hall and an ancient Aga in the kitchen provided the heating in the house. The open fireplace was really just a hole with a fire basket placed under a wide chimney. It consumed wood and coal with a voracious appetite and encouraged spine-chilling draughts from the ill-fitting Victorian windows. The house heating needed a radical overhaul.

Heating options are mains (natural) gas, electricity, oil, solid fuel or liquid petroleum gas (LPG). Mains gas is relatively cheap and is certainly convenient, but some rural areas do not have access to it. Many country properties have oil-fired central heating systems instead, with oil being delivered by tanker on a regular basis. This is expensive and likely to remain so.

Where electricity is used for heating, it is common to have an off-peak tariff where radiators heat up overnight, when it is cheaper, releasing the warmth slowly during the day.

LPG requires the installation of a pressurized tank outside the house, with a regulator for controlling pressure of gas along the pipe. A shut-off valve is also incorporated. LPG is useful, not only in remote dwellings, but also as a subsidiary form of heating.

Solid fuel includes coal, wood and smokeless fuel briquettes. The latter are widely used where clean air regulations prohibit dirty smoke emissions. All solid fuels require storage space and the availability of regular deliveries or replenishments. Wood must be allowed to dry out for at least a year before it is suitable for burning. Green wood produces tar that can build up in a chimney, causing a potential fire hazard. It also produces far more smoke than seasoned wood.

Logs need to be stacked, out of the rain where air can circulate freely. The leeward side of a dwelling with wide eaves is the traditional place for storage, but there are many alternatives, depending upon the site. One year, a pair of stoats took up residence and produced a litter of young ones in our log pile. We knew the logs must have dried out sufficiently for burning if they were deemed to be suitable housing, but we used logs from elsewhere until the stoats departed, so as not to disturb them.

Stoves

For many years woodstoves were not efficient, and the smoke given off released pollutants into the air. In recent years, their design has improved considerably, so more of the energy is converted into heat and less comes out of the chimney as smoke. Wood burns on a flat bed of ash with an air feed above, enabling combustion to take place. The gases given off cool as they rise, but the introduction of a secondary air input to ignite and burn the combustion gases produces a second burning process, providing extra heat output and cleaner emerging gases. The phenomenon is known as 'clean burning'.

A well made stove needs to be airtight. This is not a problem with welded steel as long as the doors continue to fit well after use, but cast-iron stoves need to be precision made, and the seals between parts checked as time goes on. If cool air enters

A small farm in New England with logs ready for burning, under cover. Those on the right are waiting to be cut. Pumpkins are on sale to callers. The chimney and added section of roof are for maple syrup production.

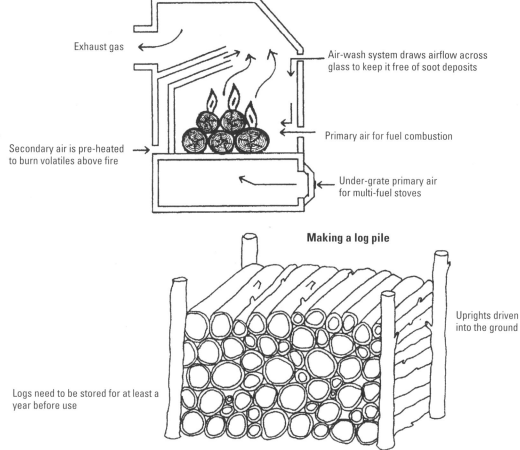

Exhaust gas

Air-wash system draws airflow across glass to keep it free of soot deposits

Primary air for fuel combustion

Secondary air is pre-heated to burn volatiles above fire

Under-grate primary air for multi-fuel stoves

Making a log pile

Uprights driven into the ground

Logs need to be stored for at least a year before use

How a stove works.

through cracks in the joints, the internal temperature drops. This also happens if a stove is run at a low combustion for a long time and there is insufficient heat to burn off the polluting particles.

Modern multi-fuel stoves are efficient and clean burning, and are a considerable improvement on earlier versions, and on wood-only burners. They meet the requirements of the most stringent clean air regulations. When choosing a multi-fuel stove, the first concern is the balance of fuels to be used. From this, it is possible to select a suitable appliance. If the stove is to include a boiler for a number of radiators, care is needed. The right-sized stove for these tasks may have too large a heat output for the room in which it is sited. For room heating only, allow 0.05kw per cubic metre.

Initially we installed a woodstove in the central hall of the house. It stood on a stone flag, projecting out so that it sent comforting glows of warmth in all directions, including up the stairs. The open fireplace in the living room was replaced by another stove, with a back-boiler. This stove was chosen because it had a glass front so that the fire could be seen, a factor we felt to be essential in the living room.

The addition of a back-boiler was to feed radiators in the identical space above, which was a large double bedroom. A local builder divided this room up into two separate rooms and a landing with store cupboard. The new rooms provided a single bedroom each for the two older children. Each had a radiator installed, connected to the woodstove back-boiler by 28mm (1in) copper pipe. The design required one high point and needed to be a complete, simple loop. In this way the heated water rose by convection and circulated through the radiators by gravity; it needed no electric pump and so was not at the mercy of winter power cuts. In later years we replaced this system with oil-fired central heating, although the stoves were retained for local heating.

Avoiding Chimney Problems

For any kind of fire, a good chimney is essential otherwise there may be problems of smoke or fumes coming into the room. The most common reason is that the fireplace opening is too large in relation to the flue size, so the volume of smoke cannot escape. The converse can also be true, where if the flue is too large, as with an old inglenook, then there may not be sufficient draught, and smoke will wander out into the room. The chimney pot may be too small in relation to the flue, or there could be an obstruction in the chimney. A poorly designed throat above the fire can slow down the smoke so that some of it drifts back into the room.

If the fireplace smokes continuously but only under certain weather conditions, then there is probably a pressure zone problem where the chimney is in the line of the prevailing wind. This may require raising the chimney. When weather conditions produce intermittent puffing of smoke or fumes, there may be a downdraught where wind passing over a high hill, building or tree, turns downwards into the chimney. This can be cured with a cowl or slab top to the chimney. If the fire smokes only when all the doors and windows are closed, then there is insufficient air for the fire, and fumes, including poisonous gases, will spill into the room: a dangerous situation. Under these circumstances there is probably nothing wrong with the fire or flue system, but it is essential to introduce more air into the room for combustion to take place!

In Britain, all houses built since 1965 have had their chimney flues lined. Houses built before then usually have rendered internal chimney walls, and it is essential to install a chimney lining in these cases because fumes or creosote from woodsmoke can leak through the chimney fabric, which is unsightly and is also a fire hazard. There are different types of liner suitable for different fuels, and advice should be taken before installing any system.

Range Cookers

There are two types of range cooker. The first is a *fixed temperature* model with either two or four ovens, each of which is set to a fixed temperature. The hob has two main hotplates: a hotter one for fast boiling, and a cooler one for simmering. The second type is the *variable temperature* cooker that usually has two ovens and a single hotplate, all of which can be manually regulated. Depending on the manufacturer, range cookers are available in a selection of models, suitable for gas (natural or LPG), electricity, oil and solid fuel (coal, seasoned wood and smokeless fuel).

All range cookers operate in basically the same way: the burning fuel heats up the metal construction of the cooker (be it pressed steel or cast iron) for the cooking facility. In the case of stoves with the dual purpose of cooking and heating, the hot stove then heats water in a surrounding water jacket for the domestic water supply and also, in some cases, for a number of central-heating radiators. The majority of ranges now available have a built-in damper control. This allows control of the rate of burning, which is important in a variable climate. This does not,

A four-oven Aga in a farmhouse kitchen. Photo: Aga

however, affect the cooking temperature, dispelling the fears many people have of a stove thundering away on a blistering summer's day to heat the bath water, but still taking six hours to cook a roast. Some models incorporate an electric element for cooking on, to meet the needs of those who prefer not to light their stove at all in summer, while some also boast a fast 'start-up' (heating up) time. The speed depends on the material used for the construction of the range, which is cast iron or pressed steel. Steel heats up and cools down quickly, whereas cast iron heats up slowly but retains its heat and cools down slowly. Some versions are fitted with catalytic linings, making cleaning easier.

Many homes with a range cooker find that the kitchen is the central point of the house, with everyone drawn to it. It is often the ideal place for drying and airing damp clothes. Logs can also be dried out next to the range, and herbs and flowers can be hung from the ceiling above to dry. Many flock owners also know the value of a range cooker when it comes to the rescue of orphan lambs.

Our kitchen was old and inefficient and the Aga, which was a warm friend in winter, was also cantankerous and unreliable: the cooker had a bad crack in it, and the thermostat did not work. A replacement choice lay between another solid fuel cooker or an electric one. Like many small villages, ours did not have mains gas available. The question of the chimney above the Aga, which had originally caused the leak in the thatch, was also a vital one. It had missing bricks, leaned to one side and was clearly unsafe. We decided to remove it entirely and cook by electricity.

Down came the chimney, felled by local builders. Everyone stood clear, but it was not such a dramatic event as a tree coming down; just quick slithers of bricks down the thatch until they landed on the grass. A few fell in the pond and caused a miniature tidal wave that set the ducks rocking from side to side in quacking protest. The old Aga proved stubborn to the end, and had to be finished off with a sledgehammer. All that remains of it now is the inner cast-iron pot from the firebox, which is still in service for the forcing of early rhubarb in the kitchen garden.

Electricity

Ideally, the outbuildings as well as the house should have electricity. If a lot of heavy machinery is to be used, it is necessary to have a three-phase electrical supply, and a qualified electrician will need to advise on, and install, the necessary equipment. The electricity company will advise on bulk and off-peak tariffs. If the wiring in the house is old, it may need to be replaced, with power points installed at specific locations. It is often a requirement of mortgage companies that rewiring be done before a mortgage is approved.

Rewiring for us was an expensive job but it was obviously an urgent priority. A local electrician installed a double 13-amp ring main with plenty of points and including a safety trip switch. The flex, bakelite switches and weird assortment of plugs became things of the past, and light came to Broad Leys. 'Just wait until the winter comes and you won't have lights', said a neighbour cryptically. And when winter came, the truth of this comment was revealed. The first high winds brought a power cut. We, like everyone else in the village, found ourselves lighting candles and waiting philosophically in the gloom. We had experienced power cuts in the town before, but there was usually a reason for them. Now we were to realize that, even at the best of times, the supply of electricity is a more nebulous thing in villages than it is in urban areas.

The kitchen is an area that uses a range of electrical appliances. We redesigned ours to meet our needs and to incorporate energy-saving appliances. Many modern electrical appliances are now designed for energy saving and environment protection: washing machines and dishwashers are energy rated, and the best models use a minimum amount of water. Refrigerators and freezers with pentane refrigerants are also available, which do not contribute to global warming. A quarter of all carbon dioxide emissions come from domestic households, so any efforts in this direction are good for the planet as well as the purse.

Lighting

Most properties will have electricity for lighting, but energy saving is possible here, too. The usual tungsten light bulb works by heating up a filament to white heat, and this is not efficient because most of the energy produced goes in heat. An energy-saving alternative is the compact fluorescent lamp (CFL), where electrical energy is supplied to an inert gas, producing an emission of ultra-violet light. This strikes the coating on the inside wall of the tube, causing it to fluoresce brightly. A CFL is not only more efficient, but also lasts much longer than a normal light bulb. It is more expensive, however, and the best savings are made in areas of frequent use such as hallways, porches, living rooms and kitchens. It is not suitable for outside security lighting which turns on and off, and it cannot be used anywhere with dimmer switches.

Emergency or subsidiary lighting is available from oil, gas, or solar-powered lamps. Rechargeable torches are also a boon to anyone who has to make frequent trips outside.

Energy-saving lightbulb.

Solar Energy

For remote houses and barns that would cost a great deal to have connected to the grid, a renewable energy source is worth considering.

Sunlight can be converted directly into electricity using photovoltaic (PV) solar panels; these are quite different from the panels that heat water. They are familiar on calculators, and from pictures of orbiting satellites where they had their first application, and in the last few years their efficiency has improved dramatically. There are also PV roof tiles available, allowing a south-facing area of a roof to be covered. The cost of PV modules is still comparatively high, although this is falling. There is not enough sunlight in northern areas such as Britain, although in the tropics, and remote areas of southern Europe, PV is used to provide electricity all year round. In areas where sunlight is limited, PV will work best as part of a renewable energy package, supplemented by wind energy or a generator during winter months. Complete solar electric kits are available, or they can be put together on a DIY basis. A system consists of four elements:

The **PV panel** that converts sunlight into direct current (DC) electricity.

The **battery** that stores electricity until needed. This should be a deep-cycling one such as a fork-lift truck battery, or a bank of nickel cadmium cells; a car battery is not suitable. A commonly used type is a lead acid battery, but there are also sealed, no-maintenance ones.

The **charge controller** that protects the battery by preventing over-charging or over-discharging; the best type has a low voltage disconnect facility. There should also be a gauge to indicate the current charge of the battery; if this is not included, a battery charger can be used.

The **appliance** or the equipment that is being powered by the system. Lighting is the most common form, but there are other applications, such as pumping water.

A PV electric system will produce 12v or 24v DC electricity, and lighting fitments such as those used in caravans are available for these. If there is sufficient power, an inverter can be installed. This converts the electrical input into 240v AC (the type normally available from the mains) so that house appliances such as a food mixer can be used.

A PV solar panel for providing electricity in a house in the south of Spain.

However, the inverter itself uses power, so there must be sufficient input from the system to make it worthwhile.

Solar Panels

A blackened surface insulated from below and with glass above will trap the heat from the sun. This heat can then be captured in water pipes, and carried away as hot water and stored for later use. This is the simplest solar water-heating panel, and such panels can be used in a domestic hot water system, or for increasing the water temperature of swimming pools. In northern areas of the world, solar panels cannot provide sufficient hot water all year round; they are normally part of a system.

A system of panels linked into a water-heating system can be purchased as a complete package; this is the most expensive option. Secondly, the panels and other parts needed can be purchased and

Solar Water-Heating Panels

Type	Comment
Flat-plate collector	Although black paint is good at absorbing sunlight, it also re-emits the heat, so panels today use a selective surface that lets out less heat. The glazing material will let in daylight but reflect back the infra-red radiation onto the panel. Although glass is good in this respect, commercial panels sometimes use special plastic materials to maximize these properties.
Evacuated tube	This is quite different from a flat-plate collector and is more efficient; it is also more expensive. It consists of a strip of material with a heat pipe that carries heat rapidly away by evaporation and condensation, so reducing heat losses. The collecting strip and tube are surrounded by a glass tube with a vacuum in between. A panel may consist of twenty or more tubes connected together.

Evacuated tube solar panel.

installed on a DIY basis, which is less expensive. Finally, the panels themselves can be constructed at home, which is the least expensive method.

The system used will depend on fuel costs. For a large household, a high efficiency system will save the most money, but needs to be balanced against the initial expense of setting it up. Cheaper systems of DIY or cheap flat-plate collectors are the other options. A typical house would be likely to need one of the following: 5sq m (54sq ft) of DIY panels, 4sq m (43sq ft) of commercial flat-plate panels, or 3sq m (31sq ft) of evacuated tube. There are a number of ways of incorporating solar panels into a water-heating system.

Thermo-Syphon

This avoids the need and cost of a pump. The method requires the panels to be placed well below the water storage tank; in this way, cold water coming down from the tank into the panel displaces, by convection, the heated water in the panel. This rises into the tank, thus effecting a slow, continuous circulation of water downwards from the bottom of the tank, through the panel, where it is heated, then back up to the top of the tank. This action continues as long as the sun shines. The pipe-work and the tank should be well insulated against heat loss. With this system, there needs to be as little horizontal distance as possible between the tank and the panel, and ideally, there should be at least 1m (3.3ft) between the top of the panel and the bottom of the tank. The pipes running to and from the panel must run downhill and uphill all the way, respectively, and ideally be 25mm (1in) in diameter. This system is the simplest and cheapest to install, although it is not possible to have the panels on the roof.

Pump System

This, as the name indicates, has a pump and small-bore piping. It may be a direct or indirect system. An indirect one keeps the water passing through the panel, separately from the hot water that is used; this solar loop passes through a heat exchanger coil in a pre-heating cylinder. With this system, the solar loop can incorporate anti-freeze (not car anti-freeze) and a corrosion inhibitor.

With a direct system, the heated water from the solar panel goes straight into the top of the water tank and down from the bottom, into the panel, as with the thermo-syphon. This is a little more efficient than an indirect system, but frost protection will be needed, which will entail draining down the panels. This can be done manually, but a safer method is an automatic, self-draining system, which can be designed with the layout.

The pump should only be on when a panel is in

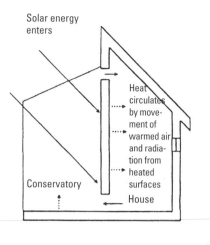

A conservatory on a south-facing wall increases solar gain.

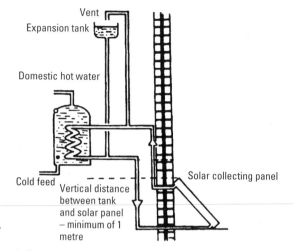

A thermo-syphoning system

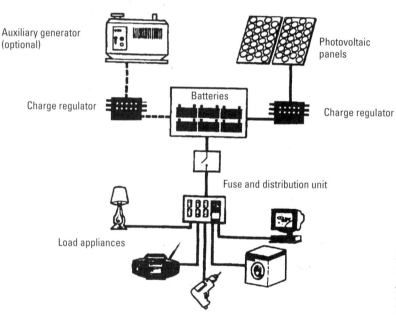

System powered by photovoltaics with an optional auxiliary generator. Diagram by courtesy of the Centre for Alternative Technology

use. It needs a solar control unit to measure the difference of the temperature of water leaving the bottom of the tank and that emerging from the panel, and it switches on and off as required. The system also needs a one-way check valve to prevent hot water flowing into the panel when the pump is off at night.

Passive Solar Energy
Although solar energy can be captured by PV or solar water-heating panels, the most effective option

in northern areas of the world has proved to be in the passive solar design of the buildings themselves. Such a design uses the form, materials and orientation of the building to capture, store and distribute the solar energy received.

To achieve maximum efficiency, a house should let in sunlight when it needs heat, and keep it out when it does not. It should also let in coolness when needed. Effective energy storage is achieved by using materials of high thermal mass in the fabric of the building, such as brick, concrete or stone, which can store heat (or coolness) over an extended period. The size and position of windows is crucial. A new house can be built using passive solar design features at little or no extra cost, compared with conventional designs. Finally, high standards of insulation, coupled with an open interior design, keep heat or coolness inside and distribute it to all rooms. In this way, the building becomes the heating and cooling system, rather than relying on additional mechanical equipment.

The addition of unheated conservatories on the south side of houses can make a considerable contribution to energy savings, if linked to controlled ventilation. Air can be drawn from the top of the conservatory into the house, using a fan if needed. An artificially heated conservatory, by contrast, may provide attractive living space, but is highly inefficient in energy terms; although for comfort's sake, this may be an acceptable compromise.

A south-facing wall can also be adapted to collect solar energy, by constructing a *trombe* wall. This is a shallow sun space between the wall, ideally blackened, and a glass outer wall. The heat from the air is absorbed by the wall during the day, and then radiated gently at night.

The final improvement we made to our house was our most ambitious project: to build an extension along the whole of the south-facing side of the house; one half was to be a dining room, while the other would be a conservatory designed to maximize the passive solar energy available. On the north side, we built a two-storey extension that provided a new living room downstairs, and a double bedroom and bathroom upstairs. Our house was finally warm, spacious and comfortable.

Wind Energy

Installing a windmill or wind turbine to generate electricity on a small scale provides almost free, pollution-free power, once the cost of equipment and installation has been met. The snag is that this cost is high in comparison with the cost of purchasing electricity from the grid, so at present it is not likely to be cost effective, unless it is required for remote dwellings and farms still unconnected. Britain is a good country for wind power, with an average wind speed of 7–8.5m/sec. Wind resource maps are available, showing what areas are suitable for utilizing wind power.

In Britain, planning permission is required for small wind turbines, but there is rarely a problem, as long as it does not impinge on a neighbour's property.

A small-scale aerogenerator suitable for house electricity. Photo: DGMS

In a remote situation a wind turbine would need to be connected with a complementary power source, usually a power generator running on diesel or LPG. This can take care of the heaviest loads, and a battery charger can be run with any spare capacity. The system may also incorporate solar power. As with PV solar panels, the wind turbine charges batteries that provide electricity in DC form for lighting and other low energy uses. If there is sufficient energy being generated, an inverter can be utilized to convert the output to 240v AC electricity so that a wider range of appliances can be used.

The first thing to consider is whether the proposed site is suitable for wind power. The power of the wind is proportional to the cube of its speed, so twice the wind speed gives eight times the power. The ideal site is a smooth hilltop with a clear opening to the prevailing wind direction. Large trees and buildings produce turbulence, increasing wear and tear and reducing the effectiveness of the wind turbine.

If the site is suitable, where should the windmill be positioned? As wind speed increases with height, it might pay to use a smaller model on a high tower. Ideally, wind speeds on the site should be measured over a matter of months to find the best position. There is a range of wind monitoring equipment available that is used before installing large grid-connected turbines. However, the cost can be considerable, and may not be worth the expense for a small installation.

There are two ways of assessing the wind speed conditions on site. First, obtain the regular wind speed information from the local airport or meteorological office. Second, set up a pole and streamers to assess the wind speed and direction, relating this to the Beaufort scale. This is a traditional method of measuring weather conditions by watching the movements of branches and rising smoke. The pole and streamers can be set up and compared with nearby trees, to see how different Beaufort numbers affect their movements; this enables the results to be calibrated. To have any validity, this exercise needs to be done over a period of months. Taking regular readings will gradually build up the picture needed. Test for height as well, because as referred to earlier, wind speed increases with height, and twice the wind speed will give eight times the power.

Small wind turbines vary in the diameter of the rotor blades. This surface area determines the potential energy available, with the smaller ones producing around 50W and the larger ones about 2kW. An important consideration is the start-up speed. Single-blade units are more efficient in high winds, but require a stronger wind speed to start up; a six-bladed unit has more torque and therefore requires a low wind speed to start up. Each site varies, and professional advice is needed when it comes to deciding upon an appropriate system. There is also the safety factor: a wind turbine can rotate at a very high velocity, and if it should become dislodged, it could cause extensive damage. Professional installation is therefore essential.

A tower enables the turbine to be placed at the correct height for the site, while a battery stores the generated power until required. This has a voltage regulator to protect it.

Power Generators

Power generators are engines run on petrol, diesel or LPG. Normally housed in an outbuilding, they can be used as an emergency source of power, or as a back-up for an alternative system. LPG generators are the least expensive and the cleanest to run. Petrol is the most expensive.

The salient points are to ensure that the generator is bought from a reputable source, and that there is a local servicing facility. The unit should have a circuit breaker to prevent overloading, as well as a switch to choose between the generator and other sources of power. The engine will need regular cleaning and servicing, and it is a good idea to run it periodically, when not otherwise in use, to keep it running smoothly.

Batteries

Car batteries are not recommended as storage units. The most cost-effective form of storing electricity is in deep-cycling lead acid batteries, or a bank of nickel cadmium 2v cells in series. The electrical energy reaching the cells is converted into chemical energy and stored until the battery is discharged. Each cell stores 2v, so a 12v battery consists of six cells, connected in series. The capacity of a battery measures how much electrical charge it can store, measured in amp hours (AH). A controller is necessary between the generator and battery to protect the latter.

It is important to keep batteries clean, away from heat, and in a protected area. Making a wooden box for them is effective. Lead acid batteries need to be

The Beaufort Wind Scale			
Force Number	Wind speed (mph)	Description	Effect
0	1	Calm, still	Smoke rises vertically.
1	1–3	Light air	Smoke drifts.
3	4–7	Light breeze	Wind felt, leaves rustle, weather vane moves.
4	13–18	Moderate breeze	Raises dust, moves twigs, opens flags.
5	19–24	Fresh breeze	Small trees in leaf sway.
6	25–31	Strong wind	Big branches move, overhead wires whistle.
7	32–38	Moderate gale	Whole trees move, some resistance to walkers.
8	39–46	Fresh gale	Breaks twigs, impedes progress.
9	47–54	Strong gale	Blows off roof tiles and chimney pots.
10	55–63	Full gale	Uproots trees.
11	64–72	Storm	Causes structural damage.
12	73–82	Hurricane	Widespread damage.

kept topped up with distilled water.

Setting up an Alternative System

To set up an alternative system it is important to plan the system as a whole, considering the power sources to be used (PV solar, solar panels, wind, generator), the cables or means of carrying the power, the control systems, and the batteries for storage. Before installing a system, one needs to calculate how much energy is needed.

Heating and cooking use large amounts of energy, and other sources such as solid fuel or LPG gas should be used for these. The system is best used for appliances that cannot be driven by other means and which use small amounts of energy. There is a range of appliances designed to run on 12v DC. As referred to earlier, 240v appliances will need an inverter that converts low voltage DC to mains voltage AC. This is an extra expense and its use involves an energy loss, but it does enable the driving of other appliances, such as a TV, computer and low-powered kitchen equipment. DC current needs thicker cables to limit the loss of power due to the drop in voltage. The longer the cable, the greater the resistance and power loss.

It is necessary to add up the appliances likely to be used, because the total equipment adds up to the load on the system; this is usually measured in watt-hours (Wh). Estimate the average number of hours each device will be in use, then multiply hours by watts to arrive at power requirements per day. For example, a 25W DC bulb that is on for four hours will require 100 Wh DC per day; or a 500W AC food mixer in use for fifteen minutes will require $500 \times 0.25\text{hr} = 125$ Wh AC per day. Finally, multiply the Wh AC load by 1.2 to allow for inverter inefficiency. Remember that energy requirements vary with the seasons, and more energy is consumed in the darker, colder winter months. Unfortunately, this is the period when solar power from PV panels is at its minimum.

To calculate the amount of energy available from PV panels, it is necessary to know the average amount of energy from the sun in the area. This is called the **insolation** level, and details are available from weather stations. In Britain, during spring for example, the rating is between thirty-six in the south to thirty-two in the north of Scotland. This is the number of kilowatt hours available per square metre per day. There are different maps for each season.

It will be necessary to generate 1.5 times the load calculated to make up for losses from inefficiencies in the system. The energy produced from the PV array in a day is equal to the number of panels × output of each panel × peak hours of sunshine.

To estimate the battery capacity needed, multiply the average daily load by the number of days reserve. This is often calculated at three days holdover time, for periods when there is little power input because of weather conditions.

A certain amount of control equipment is needed within the system. A charge regulator between the generator and the batteries protects them from being damaged: it prevents too much energy being fed into them, and disconnects the loads if the battery loses more than half its charge. When the battery is recharged, the loads are automatically reconnected. Using an ammeter, the energy that the PV panels or other generators are putting into the battery, and how much the loads are taking out, can be measured. A voltmeter can also be connected to the battery to see how much energy is stored there.

Water Supply

Water is an essential commodity, and while a private source may be adequate, most smallholdings now receive their supply from the mains source. In an old building, there may still be lead pipes, although most have now been replaced by copper piping to avoid leaching of poisonous lead into the drinking water. Some old outbuildings were not equipped with a water supply at all, and this is an important factor to bear in mind before purchase. Can the existing system be extended without too much cost? As they use more water than the domestic consumer, farmers usually have a metered supply so that they are paying for what they actually use. Because of this, it is important to check that the existing pipes and outlets are sound, for if a leak develops in a section of underground pipe, you may be paying for wasted water.

A natural spring on the site may be appropriate for domestic or livestock use. If there is a continuously flowing stream, it could provide hydro-electricity for remote buildings. There are companies that specialize in this. Different countries have their own regulations. In Britain, for example, it is necessary to have a licence from the Environment Agency in order to extract water from a spring, borehole or river. A good stream will provide power round the clock so that the electricity can be used directly, or it can be stored in batteries. A solid state, electronic load controller would enable any excess power to be used in water or space heating.

With a private supply, it is important to establish

Water is an essential commodity, but a mains supply may not extend to livestock areas.

that there is sufficient flow throughout the year. If it is higher than the house, then gravity will deliver it, otherwise a pump will be needed to fill a tank that is placed above the building. It is then relatively simple to buy or construct a filter for filtering out impurities from a private water supply. There are a number of ways of transferring a continuous flow of water uphill, including pumps, hydraulic rams, and wind and water systems.

Pumps

The majority of pumps used in small water supplies are powered by electricity. If there is no convenient electrical supply, a diesel engine can be used, but they are noisy and use a lot of fuel. Pumps work by lifting water to the pump intake, then applying pressure to the water in the pump to force it through the delivery pipe. They can be classified into constant displacement pumps (CDP) and variable displacement pumps (VDP): these deliver the same

quantity of water for a wide range of head. The power required to drive the pump is determined by the height to which the water needs to be pumped.

There are three types of CDP. The first is the reciprocating piston pump that uses the movement of a plunger to displace water in a cylinder. An example is the traditional hand pump, where the pumping rate can be varied by changing the speed of the plunger or piston. They are relatively inexpensive and easy to maintain. The second CDP is the rotary pump that uses rotating vanes to suck in water and force it out on the other side. These are normally surface pumps, and may suffer wear and tear from sand and grit being sucked into the vanes. The third CDP is the helical rotor pump, which is a modified rotary pump. It can be used at the surface or in a deep well, and is driven by a polished metal rotor. Sand and grit do not have a damaging effect on it.

VDPs can be divided into two types. The more common is the centrifugal pump that has a rotating impeller and uses centrifugal force to pump the water. There are a number of designs for this type, and they can be used on the surface or for deep well applications. The remaining type is the jet pump which is only used in wells and boreholes. They are cheap to buy and maintain, but are inefficient and use a lot of power. However, they are adaptable for even the smallest wells, and are simple to operate and maintain.

Hydraulic Ram

This is an alternative form of pump that requires no outside power input and has no moving metal parts. A ram works by using the momentum of a relatively large flow of water under a small head to raise a small quantity of water against a large head. Water flows from a supply stream or spring through the pulse valve that is shaped like a mushroom. As the water flow increases, it closes the valve, which then causes a pressure rise in the chamber. This opens a second valve, sending water up the delivery tube. As this happens, the water pressure in the ram falls until it is too low to keep the delivery valve open. It then closes, and the reduced pressure then enables the pulse valve to open again, and so on.

Hydraulic rams can be made to work with a head as low as 50cm (20in) with a volume input of between 4–2,000l (1–445gal) per minute. The lift potential is up to thirty times the 'fall', so a 1m (39in) fall will give a maximum lift of 30m (98ft). Rams are ruggedly built and require very little maintenance apart from occasionally replacing the rubber parts of the valves.

A close-bladed windmill suitable for pumping water.

Wind and Sun Water Systems

Multiple-vane windmills can be used in conjunction with a water pump. They are normally installed on a steel tower about 15–20m (50–65ft) high. They require a wind speed of at least 10kmph and normal wind speeds of 25–40kmph for optimum operation. PV panels that convert solar energy into electricity can be used in conjunction with the system. Wind pumps work only when there is sufficient wind, and PV panels when there is enough sun. Hopefully, they complement each other.

A Filtration Unit

With a water filtration system, water is piped into a filter equipped with a valve control, and then percolates down through sand and pea gravel, which strains out the impurities. A ball valve in the filter tank maintains the water level. Although such a system can be successful, it does not guarantee that *all* harmful bacteria are removed. For this, an on-line particle filter and an ultra-violet filter should be incorporated in order to kill any remaining bacteria.

A private source such as a well or spring needs to be checked for purity on a regular basis. Dairy farms must have their water supply tested for purity, regardless of its source, to ensure that the regulations relating to hygienic milk production are met.

Off-Mains Mains Drainage Systems

If the property is not connected to the mains drainage system, the problem of dealing with waste will be the householder's. For effective and legal disposal, a septic tank or a mini-treatment plant will be required. If neither is present, planning permission will be required in order to install a system. In Britain, this involves securing a 'Consent to Discharge' from the Environment Agency. It is essential to obtain professional advice, to ensure that the system is suitable for the number of people in the household, as well as the rate of percolation through the soak-away.

Septic Tanks

A septic tank is the simplest and most economic solution, since it provides a degree of treatment

A septic tank being installed. Photo: Conder

before discharging via a soak-away to the subsoil. Further bacteriological action then takes place ultimately to render the effluent harmless. A typical tank will be about 3m (10ft) high, generally spherical, and will hold around 3,000l (825gal). It must be installed below ground, a minimum of 15m (50ft) from the property. All household liquid waste, except roof and surface water, is fed into the tank. Internal baffles help settle out solid matter, leaving the liquid effluent to be discharged safely into a soak-away or sub-surface drainage system. This effluent can still be quite polluting and, therefore, cannot be discharged into a watercourse. The settled solids are retained in the tank and decompose into a sludge that needs to be removed annually by tanker.

Septic tanks operate efficiently, as long as there is not a build-up of detergents, soaps and fatty substances that can destroy the bacterial action, causing blockages and foul smells. Proprietary products are available to keep the systems working properly by encouraging the necessary bacteria.

Mini-Treatment Plants

With the focus growing on higher levels of environmental protection, a septic tank may not always be appropriate. In this situation, the most likely solution would be a small packaged treatment plant. This is a comparatively recent development, and is essentially a system with three treatment zones within a single tank design. There are a number of systems on the market that operate in a variety of ways, but without exception they seek to create an artificial environment in which oxygen levels are increased to accelerate the biological activity in the effluent.

As with an effective septic tank, the mini-plants separate the solids, but in addition, treat the effluent mechanically. This encourages the biological process and produces a high quality liquid discharge. The solids are removed by tanker at intervals recommended by the manufacturer, usually annually. The fully treated effluent can often be discharged with the consent of the Environment Agency, into a suitable watercourse. However, it is quite common to see a packaged treatment plant used with a soak-away, especially in areas where the land is not ideal for the disposal of poor quality effluent. The process is so thorough that it is likely to meet the most stringent consent conditions, where levels of nitrate may be stipulated.

The moral of the story is not to be put off that idyllic rural site just because it has no mains drainage, because providing a private treatment system is not nearly as difficult and costly as some people would have one believe; also, the perceived inconvenience is often reflected in the price of the plot.

Reed Beds

In recent years, reed beds have become popular as an efficient and aesthetically acceptable method of dealing with effluent and waste water. The first ones were installed in 1986 in Britain, and have proved effective where previously there were problems. (New regulations introduced to protect rivers and water courses have resulted in prosecutions being brought against some farmers who were polluting water courses with slurry or manure run-off from fields.)

A reed bed is essentially an area set aside to hold water; this is then filled with layers of sand, gravel and some soil. It is planted with suitable plants that protect, stabilize and insulate the bed, as well as providing aesthetic appeal. Suitable plants include common reed, *Phragmites australis*; yellow flag, *Iris pseudacorus*; bulrush, *Schoenoplectus lacustris*; and reedmace, *Typha latifolia*.

As waste water is introduced into the reed bed, it percolates down through the layers of sand and gravel which act as fine filters. The organic material retained forms a slime that is then broken down by bacterial action. The plants hide all this.

There are two types of system: horizontal flow reed beds (HFRB) and vertical flow reed beds (VFRB). The former has water flowing horizontally down a sloped base, leading down to a final discharge point. It is appropriate for smaller quantities of waste water, but as this is retained for longer periods, removal of fine particles is very effective. VFBRs are more commonly used as part of a system where larger solids have already been removed, and where the final output is then directed to a sediment bed for further treatment. They are deeper than HFRBs, cater for an increased volume, and usually require aeration pipes and pumps. There are specialist suppliers who will construct and maintain reed beds. A DIY system is also feasible for those of a practical nature, but expert advice is essential, as different sites require different treatment. Practical courses are available, and attendance at one of these is recommended.

3 Outbuildings, Machinery and Tools

If outbuildings are structurally sound, they can be adapted in various ways.

Outbuildings are essential on any smallholding. They are needed for housing livestock, storing hay, straw, feedstuffs, vehicles and equipment and, in some cases, may have a specialized function such as providing workshop or dairying facilities, or even a home office. Anyone considering buying a small-holding would be well advised to concentrate on one that has existing outbuildings in a good state of repair.

Traditional farms in Britain often have stone outbuildings. These may be lovely in appearance, but there are disadvantages. A notable one is that access for mechanized operations such as muck clearing may be difficult. A front-end loader attachment on a tractor would be impossible to use, so clearance by hand would be the only option, unless part of the wall were dismantled. If there is a preservation order on the outbuilding, this may not be possible.

Outbuildings may be cold and draughty, although a false ceiling inserted in the roof space of an A-shaped building makes the whole structure much warmer. Insulation boarding sheets can be used to line stone walls. If a new concrete floor is to be laid, a layer of insulation underneath makes a big difference. The final concrete layer can then be painted with a non-slip paint that also avoids the dust normally associated with concrete floors. This is particularly important in an outbuilding such as a milking parlour where a high degree of cleanliness is required. Here, floors and walls need to be sealed for ease of washing. We were fortunate in that one outbuilding had a concrete floor and a drain, so we decided that this would make a suitable milking area for our goats.

In some cases, grants may be available for converting traditional farm buildings for new rural enterprises, as long as the essential architectural features are conserved. In fact, making changes to the inside is not usually a problem; it is the outside that must remain the same. A friend of mine, for example, received a grant to convert an old stone barn into a shop selling traditional woollen products made from locally spun fabrics.

Ordinary garden sheds are widely available and easily erected or subsequently moved. They can be adapted for a wide range of uses, from tool store or workshop, to poultry or other small-scale housing. The range of animal and poultry housing is a wide one, with an equally wide range of materials used in their construction. Information on specific housing and specialized buildings such as a farm shop or home office are to be found in the appropriate chapters. The following buildings or areas are those that cater for more general use.

The Farmyard

It is often said that the centre of a small farm is the farmyard. Traditionally this was a yarded area bordered by outbuildings, where animals were brought for over-wintering or other activities. Having a paved or concreted area provides some-where that is free of mud (often an unforeseen aspect of living in the country). It is a place where tasks such as animal foot trimming and exercising can take place in relative comfort.

An outbuilding adapted for pigs. The concrete run has been 'combed' to prevent slipping.

In the absence of a farmyard, it is useful to have a concreted or flagstoned area outside the outbuildings.

A Dutch barn where hay and straw are stored. Here, it is also providing shelter for Muscovy ducks.

If there is no yard, it is a good idea to make a concreted or paved area near the outbuildings. If required, a non-slip coating can be added, such as that normally used inside buildings. If it is just plain concrete, it is worth 'combing' the surface in order to provide slight ridges. This stops animals slipping, a factor that is particularly important where there is a slope. A small cement mixer is a valuable addition to the smallholding, for it speeds up the whole process of making concrete. It can be bought or hired.

Hay and Straw Storage

Storage for hay and straw is essential for anyone keeping livestock. Hay and barley straw are eaten by farm animals, while straw also provides warm bedding. It is important that they have dry, airy conditions such as those provided by the traditional Dutch barn with its roof and open sides. Hay, in particular, is dangerous if it becomes damp. It can, on rare occasions, ignite spontaneously because of the high temperature brought about by decomposition. If it is mouldy, it may give off spores of a fungus causing the disease aspergillosis. This affects not only livestock, but also humans who may handle the hay. The alternative name for the disease is 'farmer's lung'.

The traditional Dutch barn is so named because it was introduced from the Netherlands in the nineteenth century, to replace the traditional hayrick which was vulnerable to the weather. It is of sheet metal construction, designed to keep hay and straw airy and free of moisture. It has one or more sides open to the air. In protected areas, all four sides may be absent, with the structure being merely four legs and a roof. In some parts of Britain, this type was designed to have a roof that could be progressively raised as the bales were added. In exposed areas, the barn may be a three-walled construction, with the open side away from prevailing winds. Wooden barns are also common in many areas of Britain and the USA.

If there is no barn on the site, an outbuilding such as a shed or stable can be used for storing straw and hay. Some outbuildings may need to have more ventilation provided, and it is a good idea to place the bottom layer of bales on a raft of planks clear of the ground so that air can circulate, reducing the risk of rising damp.

Feed Storage

A building that is free of damp is also essential for the storage of animal feeds. Equally important is the need to ensure that it is vermin proof. The depredations of rats and mice are legendary, but rats are also carriers of disease, a fact that makes them doubly dangerous.

Traditional farm storage buildings were well clear of the ground, resting on mushroom-shaped supports that prevented rodent access. Having a high enough space underneath also allowed farm cats to gain access.

If areas such as the base of doors, or the corners of buildings have been gnawed, they can be reinforced with sheet metal or rigid, galvanized wire mesh.

Field Shelters

In northern climates, grazing animals usually have permanent housing where they are taken to sleep at night. They often have additional temporary shelters in their grazing areas – though in some cases, these field shelters may be the permanent houses. Larger field shelters are usually three-sided buildings, allowing animals quick and easy access in the event of bad weather. They are normally made of timber or

A traditional feed storage building, well clear of the ground and resting on supports that prevent rats getting in.

An open-fronted field shelter is ideal for animals out in the fields. Note the hayracks at the back, and the angle of the roof to shed rainwater away from the entrance.

galvanized metal. There are also mobile structures made of wood, metal or plastic, catering for a range of different livestock.

It is quite common for polythene tunnel greenhouses to be used as animal shelters, or as lambing areas. The sides are boarded, with appropriate ventilation, while shading prevents excessive solar heating. Temporary field shelters are easily made from straw bales with a few stakes to hold them in place.

The pole barn is an adaptation of the field shelter. It is so-called because the original ones had a framework made of rounded poles set into the ground and supporting the roof and floor. The structure is widely used for animal or turkey housing, or indeed anywhere where sheltered, airy conditions are required. DIY plans are available for building barns, pole barns and animal shelters.

Machinery, Tools and Equipment

It is said that there is a tool for every task. The ingenuity of man certainly bears this out, for over the ages an extraordinary range has been developed. Some tools are even adapted for local conditions and working traditions, as the range of sickles and billhooks in different regions illustrates. Master craftsmen frequently make or adapt their tools in order to fit their own hands.

A smallholding may already have a collection of old agricultural vehicles and tools, although this is becoming rare now that collectors have realized their value. Where some machinery is concerned, it may be more appropriate to sell to a collector and use the money to buy newer equipment. The first question, however, is to decide on the scale and type of operations. The second is to assess what costs and skills are involved. Those using tractors, ploughs and other machinery would be well advised to attend one of the many courses offered by agricultural colleges. The art of welding is particularly useful for those who wish to carry out their own repairs. Again, courses are available which cover electric arc and metal inert gas (MIG) welding.

The safety aspect is also crucial. Vehicles, machinery and tools need to be stored safely and securely where they do not present a hazard to children and adults alike. This includes having locked storerooms as well as the appropriate safety wear when using them.

Time and energy can be saved using appropriate machinery for different tasks. Some machines are versatile and will tackle a number of different jobs. It is unlikely that anyone would want to have the whole range of machinery, tools and equipment available for smallholding use. It is better to have a few good basic tools that can be used in a variety of situations, and acquire specialized equipment as and when required. Furthermore, it is not necessary to buy such equipment, for many items can now be hired. There are also frequent sales of second-hand vehicles, machinery and tools, with some companies specializing in this area.

Some tasks, such as ditching, drainage, tree surgery and extensive hedge cutting may be better left to a contractor who will come in with his own equipment and do the whole operation in a short time. There are also tasks that involve the use of potentially dangerous tools. These may be better left to an expert, unless a practical course in the use and maintenance of such equipment has been undertaken.

Tools, machinery and equipment that are required for specific tasks are described in the appropriate chapters of the book. The following are those that are appropriate for more general use.

Clearing Rough Ground

Most smallholdings have some rough areas where long grasses and brambles proliferate, and this is where a strimmer or a brush cutter can prove useful. A strimmer uses a fast-rotating nylon line and is suitable for grasses and relatively soft stalks. A brush cutter has a range of cutting blades, making it more appropriate for brushwood. Lightweight strimmers are powered by electricity, while brush cutters are petrol driven. Engines vary in size, but are usually two-stroke for lightness. Four-stroke ones are available, but these are generally mounted on wheels; they are often used in orchards.

Eye and ear protection is advisable, as well as stout boots. It is often forgotten, or not appreciated, that back injury may result after prolonged use. The machine needs to be well balanced. If it has a harness, this should be fitted properly for the individual operator.

Nylon lines and replacement blades are available for strimmers and brush cutters. The latter can use a range of cutting heads, depending on the size and strength of the brushwood to be cut. With careful use, avoiding stones and other obstacles, these machines will give good service for many years, as long as they are kept clean, oiled and serviced where necessary.

Breaking the Ground

Walk-along machine cultivators or two-wheeled tractors are the most appropriate tools for the large garden or small acreage. They are available in a range of 5 to 20hp. Some growers use a large machine for initial ground clearance, then a smaller one for producing a fine tilth or for inter-row cultivation. We bought a small cultivator and hired a large one when necessary. Small models are cheaper, more manoeuvrable and easier to operate and repair. There is a wide range of cultivators available, with an equally large selection of fitments to clear, plough, harrow and generally cultivate the ground.

There are essentially three parts of a cultivator to be considered: the motor, the drive and the blades.

A brush cutter is useful for overgrown areas. Photo: Stihl

A walk-along cultivator with a range of attachments will do most of the land tasks on a small acreage. Photo: Agria

The motor needs to be kept clean, with its spark plug checked and oil changed as necessary. At the end of the growing season it should be checked over, and the tank should be drained before it is stored in a dry area. On older machines there may be a belt drive from the motor, while newer models often have chain drives. The former should be checked for wear and if necessary replaced, while the latter should be well oiled. The blades are normally sturdy and hard wearing, but may sustain damage from large stones. They are usually replaceable.

Many problems can be avoided by keeping the cultivator clean, and having a replacement spark plug to hand if there is a starting problem. If petrol is put into the tank through a funnel with a filter, and the top of the tank kept clean, there is less likelihood of dirt getting in and blocking the carburettor.

For a larger area, a four-wheel tractor will be useful. It is the general workhorse of the farm, and a power take-off (PTO) facility will enable a wide range of implements to be used. A three-point hydraulic lift linkage will also be required in order to raise and lower implements, as, for example, at the end of a cultivation row.

The second-hand Fergie (Ferguson) is popular with smallholders, and there is an active trade in old models that can be repaired and upgraded. With attachments such as plough, harrow, seeder and flail cutter, it will do much of the ground preparation and crop planting. With additional machinery such as a grass scythe, swathe turner and baler attachment it can be used for haymaking. One of the great advantages of the old Ferguson was the cheap availability of implements, although these are now becoming more difficult to find. Local blacksmiths will often help with the production or repair of implements. Most agricultural shows have machinery sections where it is possible to view second-hand vehicles and machinery, as well as talk to those who are knowledgeable about their maintenance.

Older tractors are fairly straightforward to maintain. The oil and coolant levels need to be checked periodically, as well as the tyres and the battery. They are inexpensive to run, for agricultural (red) diesel is considerably cheaper than normal diesel.

In recent years, an interesting development has been the availability of small tractors that fill the gap that used to exist between the garden-sized tractor and the full-sized agricultural one. Most of these are from Eastern Europe or Asia, and are at more affordable prices. Some are already finding their way into the second-hand market.

All terrain vehicles (ATVs) are popular, particularly for uneven ground, and are effective at moving loads with the addition of a trailer.

Moving Heavy or Bulky Loads

There are some excellent wheelbarrows that are well balanced and can be used to carry loads with ease. Those with a polypropylene base are much lighter than metal or traditional wooden ones. Some can be fitted with extensions to increase the carrying capacity, but these are normally for light materials such as grass and leaves. We managed for some time with a normal wheelbarrow, but eventually acquired an American farm cart with a large load area and big wheels. This could carry six hay bales or eight straw bales in one journey, with no fatigue.

Heavy loads, or moving things over an extended

Tractor ploughing a walled garden area prior to harrowing and cultivating. Note the roll-over bar for safety.

A deep, lightweight barrow can be used to transport a range of things on the smallholding, including hay bales and feed sacks. Photo: The Big Barrow Company

A covered trailer with a non-slip ramp is essential for transporting livestock.

distance, can be a strain. Power barrows will carry from 250–1,500kg (5–30cwt) and can be tipped for emptying. Some are ride-on models, while others have folding handles, enabling the user to walk alongside if necessary.

For a tractor or ATV owner, there is a range of trailers available for moving loads, which may or may not have a cover. If animals are to be transported to agricultural shows or to markets, the trailer needs to be a covered one, and to have a non-slip ramp for access. Livestock trailers are available in different sizes, and ensure that the animals do not come to harm in transit. A tow-bar is required for the towing vehicle, and for the trailer, a clearly visible vehicle registration plate, and rear, brake and reversing lights.

Cutting and Splitting Wood

A good chainsaw makes light work of long, heavy jobs such as tree felling and cutting up logs for firewood. However, it is a tool that should only be used by skilled hands, and a practical course of instruction is essential. A sawhorse that enables a log to be clamped into position before cutting is advisable.

There is a considerable range of chainsaws available, but for general smallholding work, one with an engine within the range of 2.8–4.0hp, and a guide bar of 36–50cm (14–20in) would normally be sufficient, powered by rechargeable batteries or petrol-driven motor, the latter being more appropriate for heavy use. The former is adequate for relatively light or occasional use. Power models are also available to run on lead-free petrol. Whatever type is chosen, the machine should be equipped with certain safety features. 'Kick-back' protection is vital: if, for example, the saw comes into contact with a knot in the branch, it may make a

sudden, upward movement, and the kick-back feature stops the saw immediately to protect the user. The chain brake should also have a lever to stop the chain while the engine is still running and the saw is not in use. Hand-guards and anti-vibration handles are also important.

Most chainsaws are supplied with a manual and a tool kit, including a chain-sharpening file and a plug spanner. A useful addition is a file holder with sharpening guide that shows the appropriate filing angles. A chainsaw that is producing a lot of sawdust, rather than wood chips, is blunt and in need of sharpening. There should be an air filtration system to stop the engine becoming clogged with sawdust. Protective clothing is vital, and for commercial use is required by law. It includes reinforced overalls or trousers, chainsaw gloves, and a helmet with eye visor and ear protectors. Steel-tipped boots are also required.

Splitting broad logs down into manageable widths for the fire can be done with a splitting axe that has a wedge-shaped blade, or by driving a metal wedge into the wood with a sledge hammer. Where there is a large quantity to get through, and a tractor with PTO is available, there are also 25–50hp mechanical log splitters available.

Protective clothing and a saw-horse make cutting logs with a chainsaw safer and easier. Photo: Stihl

A useful addition is a shredder that can be used to shred surplus twigs and prunings. On a fine setting it is excellent for producing a mulch for use in the garden; on a coarse setting, and used with wood rather then leafy material, it produces woodchips, which are useful in outside winter runs for poultry. Garden-sized shredders are run off electricity, and as with all electrical appliances used outside, a safety cut-out plug should be used. Larger models are available, including those that can be run off a tractor.

There are many other items of equipment that are useful on a small farm, including workshop tools, gardening tools and miscellaneous items. The table (*see* next page) is a general guide to the different farm machinery, and hopefully will help you decide what to buy and what to hire. Only basic farm machinery is included in the table, with priority given to that likely to be used by the small farmer. A visit to the local showroom of a farm or garden machinery company will enable the prospective buyer to see most of what is available. Agricultural and garden shows will also have a great variety on display.

Log splitting with a purpose-made splitting axe.

A shredder is ideal for producing mulch from tree prunings. This one is being examined by the author at Rodale organic farm in the USA. Note the handy cart.

Security

It is a sad fact of life that crime in the countryside is rising. Thieves are mobile, and literally nowhere is totally secure today – although it should be remembered that statistically, most people will never be burgled. However, it is sensible to take basic precautions, and to make the house, land and livestock secure.

A solid wall without protrusions that might be used for climbing is the best perimeter around a property, but this is too expensive for anything other than a small garden area. A wooden fence is less strong, but the post and rail-type garden fencing with overlapping boards is quite strong and hard to climb. Walls and fences can be extended with trellis for climbing plants, such as roses, whose thorns will also act as a deterrent. For other boundaries, particularly longer ones around land, plant dense, prickly hedging against wire-mesh fencing.

Items such as ladders and garden or workshop tools should be locked inside a secure building. A garden spade can be used to lever open doors or windows. Fit window locks!

External lights can be used around the house and outbuildings; you should aim to light up all possible approaches to the property. The lighting can be adapted to switch on when someone approaches, by fitting an internal controller to set how long the light remains on after it has been triggered. It is operated by passive infra-red (PIR) sensors that detect movement. These have a photocell, to prevent switching on during daylight. Powerful lights can be fitted high up, out of reach, on buildings where they illuminate a large area. One halogen flood lamp can illuminate the equivalent of one small garden.

To protect large open areas, some form of surveillance equipment may be needed. An infra-red beam system can be installed that will not be triggered

Smallholding Machinery and Tools

Machinery	Main uses	Comment
Tractor	As a general workhorse. Ploughing, cultivating.	Smaller sites: two-wheel garden model. For larger acreage: four-wheel farm tractor.
PTO (power take-off)	Powers range of attachments.	Essential for farm tractor.
Trailer	Used for carrying animals and wide range of materials.	Essential accessory to the tractor or other vehicle with tow-bar. Smaller sites:barrow, hand-trailer or power barrow.
ATV (all terrain vehicle)	For pulling or carrying loads.	Useful on uneven ground.
Manure spreader	Trailer attachment for tractor with shredder to chop and spread natural manure onto the land.	Particularly useful for organic farmers. Alternative is to stack manure, let it rot, then spread by hand from a trailer. Important not to pollute water courses and streams.
Grass-cutter	Pulled by a tractor, cuts grass for hay and 'tops' pasture.	Relatively inexpensive. Also useful for clearing rough pastures. Smaller sites: garden mower.
Swathe-turner	Turns hay in order to dry.	If hay not turned with this attachment, it will be necessary to turn by hand with a hay rake if own hay required.
Baler	Bales cut and turned hay.	Makes hay storage easier and more economical than loose hay.
Plough	Many different kinds depending upon cultivation.	If scale warrants it, ploughing can also be done by contractor.
Harrow	Breaking up clods in land prior to sowing.	Ploughing and harrowing on a garden scale can be done by a two-wheeled cultivator.
Seed drill	Wheeled hopper towed behind tractor, for distributing seeds through evenly spaced tubes.	Only necessary on a field scale. Small, wheeled sowers available for garden scale, or can be done by hand.
Strimmer/brushcutter	Clearing overgrown land.	Hand- or wheel-driven models.
Shredder	Shredding wood and prunings.	Good for mulch and production of wood chips for winter poultry runs.
Chainsaw	Tree surgery, wood cutting.	Safer to use contractor. Practical course and safety procedures vital.
Flame gun	Clearing area for cultivation.	Useful for organic growers. Hand or tractor-mounted models.
Wood-cutting tools	Axe, log splitter, wedges.	Power models available for tractor.
Hedging tools	Billhook, shears, trimmer.	Can be done by contractor. Safety procedures vital for power trimmers.

by cats, dogs or birds, and which can be linked to a combination of lights and alarms as required. A radio beam transmitter can pick up intruders and trigger an alarm in the house some distance away, without the intruders being aware that they have been detected. Remote surveillance systems can also be linked to closed circuit TV cameras, if needed. Intruders often come onto a property to 'have a look round' before returning later to rob the premises. There are many professional firms that install systems. Whatever is used, make sure that the installation and maintenance conform to nationally recognized standards such as the British Standard BS4737.

Many areas have 'neighbourhood watch' schemes where people co-operate to their mutual benefit. The local police force provides help and information, and actively encourages neighbours to start their own schemes. Valuable items, both inside and outside the house, can be labelled with the postcode and house number. They can be marked invisibly with a security pen that can be seen only with ultra-violet light, or visibly by engraving, etching or stamping.

Finally, mention should be made of dogs that may be fine guards in addition to being family pets. Geese are also effective at deterring intruders, while guinea fowl with their sharp eyesight and strident calls make excellent burglar alarms. Call ducks, for all their small size, produce a quacking of considerable decibels.

Part 2

The Land

4 Boundaries

Thou hast set them their bounds beyond which they shall not pass. (Psalm 54)

Boundaries are physical demarcation lines that separate one area from another. They are made from a wide range of materials or plants, and include hedges, ditches, banks and fences. When buying a country property it is a good idea to walk around the whole site, checking on the position and condition of all the boundaries. Hopefully, a solicitor's searches will have uncovered any problem areas such as shared access or other rights of way, as well as who is responsible for their upkeep.

Our boundaries were a combination of natural mixed hedge, post and rail and netting, with the odd piece of galvanized metal blocking up gaps. We gradually replaced these with hedge shrubs. For selective grazing, we used portable electric fencing. We were particularly anxious to ensure that our stock did not wander off our land and encroach on any neighbouring territory, especially as one neighbour had a beautiful 3-acre rose garden, a kind of Elysian field as far as goats are concerned. The old adage of 'good fences make good neighbours' is very true.

Hedges

Hedges have many uses, including the marking of property boundaries, stock confinement and wind protection. A network of hedges provides a haven for wildlife deprived of other natural habitats. It offers food and shelter to a great variety of species and enables the relatively safe passage of wild plants, insects, birds and animals. As a source of hedgerow fruits such as blackberries, elder, juniper, mulberries and hazel nuts, hedges provide an autumn harvest for people as well as for animals and birds. They provide barriers against trespassers and, finally, are an undoubted enhancement of the landscape.

Sometimes a hedge may be in such poor condition, and with so many gaps, that it is appropriate to plant new shrubs in the spaces. There may also be occasions when a completely new hedge is planned. Planting can be carried out at any time between autumn and late winter, but it is best to avoid really frosty days. Avoid planting willows and poplars near houses if you value your foundations! Saplings are available from suppliers in various forms.

Bare-rooted saplings: These are plants that are not growing in containers and consequently need to be planted immediately. If a delay is inevitable, they

A well established hedge provides a haven for wildlife as well as making an effective barrier.

should be 'heeled in' temporarily. The roots need to be well soaked before planting, and any damaged roots should be trimmed. Bare-rooted trees and shrubs are normally the cheapest.

Cell-blocks: These are young plants grown in soil 'plugs'. Consequently there is less disturbance to the root ball and less possibility of a setback when planted.

Container trees: These represent the most expensive way of buying trees and shrubs because they will be growing in pots. They are available in different sizes – right up to mature trees – with the price reflecting the age of the plant. There are some shrubs, such as holly, which should always be bought as pot plants because they resent root disturbance.

To plant a new hedge, prepare a trench about 40cm (16in) deep within a cleared area of around 1m (3.3ft) wide. For a thick hedge, plant two staggered rows, with rows 20cm (8in) apart. The previous soil mark on the sapling is a guideline to where the soil needs to come to on the trunk. If young cell-block saplings are planted, make sure that the top of the plug just below the soil surface is covered by a thin layer of soil; this helps to ensure that the root plug does not dry out. Water the plants well and cut down to around 15cm (6in) above ground: this stimulates new growth and makes the hedge bushy at the bottom. Obviously this is a recommendation for hedging plants in general; specific plants may need different treatment, so follow the instructions provided by the supplier.

It is essential to stop weeds or grass moving into the hedge area. They compete for water and

nutrients and will have a detrimental effect on the development of the plants. The first three years in particular are when they are at their most vulnerable. Use a mulch such as wood chippings to stop weed growth and to conserve moisture. Black plastic sheeting pegged down also works well. In fact, a very effective method is to lay down a roll of this at planting time, and make holes for the saplings.

If livestock are to have access to the area where the new hedge is situated, it must be protected against them while it is in the early, vulnerable stages. Newly planted trees and shrubs may also be at risk from wild animals: rabbits and even deer may invade the garden, field or orchard, and it may be necessary to use tree-guard protectors to stop the bark being nibbled.

The choice of hedging plants will be largely determined by the main purpose of the hedge, and the nature of the soil. A good stock-proof hedge, for example, would be composed of hawthorn and blackthorn, while really damp or boggy ground would support willows, alder and buckthorn. Fruit growers avoid plants of the *Rosaceae* family such as blackthorn, crab apple, hawthorn and dog rose because of diseases such as fireblight that they have in common with apple and pear trees. Beech, dogwood and hazel do well on chalky soils.

Maintenance of Hedges

Hedges are sharp, thorny places, and stout clothes, boots and gloves are needed. Protective goggles are also a good idea, and indeed essential if a power hedge trimmer is used. Any electrical appliance should have a power breaker adaptor in the circuit, to avoid the danger of electrocution if the cable is cut.

Lopping shears will deal with bigger branches. Some of these have long handles in order to reach higher branches. A good pruning saw is necessary for the really thick branches; this narrows towards its tip, making it easier to gain access. Where large branches are removed from specimen trees, ensure that the cut is smooth and at an angle where water will be shed. Paint the wound with a proprietary pruning compound.

Existing hedges normally need only be trimmed in the latter part of the year, after the nesting period of wild birds is over. The trimming stops the hedge becoming too tall and straggly, with gaps lower down. The cutting encourages new growth from the bottom so that the whole hedge thickens. Aim for the finished boundary to be A-shaped, tapering to the top. An established evergreen hedge normally has an all-over trim once a year, with hand shears or a

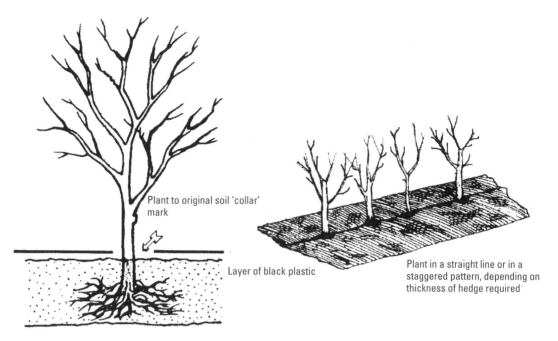

Plant to original soil 'collar' mark

Layer of black plastic

Plant in a straight line or in a staggered pattern, depending on thickness of hedge required

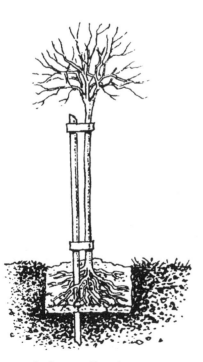

Single or specimuin tress will need supports to prevent wind disturbance. In areas where rabbits or deer are prevalent, bark guards may also be necessary.

Tree and hedge planting.

Laying a hedge

Billhook

Hedge plants partially cut through and bent over sideways. Here, they are suppored by wooden stakes hammered in at the opposite angle.

Hedging Plants

Hawthorn, *Crataegus monogyna*	Excellent, quick-growing thorny shrub. Best shrub for structure, so use it for about half your hedge. Happy in a wide variety of soils, including windswept coastal sites. Commonly planted for thick cover, its berries are a vital food supply for birds in winter.
Blackthorn, *Prunus spinosa*	Slower-growing than hawthorn but more spiny. Excellent for keeping in stock or deterring intruders. Not fussy about soil type, and trims well when established, growing many suckers from the base. Sloes are produced in autumn after the white flowers, but beware of the thorns which can produce festering wounds.
Field maple, *Acer campestre*	Fast-growing and very hardy species that can become a small tree unless cut back. When regularly pruned makes a thick branching hedge. Good autumn colour. Does well on most soils if they are reasonably well drained.
Holly, *Ilex aquifolium*	Slow-growing, but tolerates shade and is well suited to filling gaps in hedges. Spiny leaves make it a good choice for stock control, but it is poisonous. To produce Christmas berries it is necessary to have both 'male' and 'female' shrubs.
Hazel, *Corylis avellana*	Strong hedging plant on drier soils. Needs frequent trimming to encourage branching. Good source of pea sticks! Quick-growing with catkins in spring, followed by nuts in autumn. For a good crop of nuts, plant different varieties such as *Kentish cob* and *purple-leaved filbert*.
Hornbeam, *Carpinus betulus*	Grows under shade (though slowly), and retains its leaves over the winter if trimmed like beech. Does best on fairly heavy soils, so avoid it if you have thin, sandy or other light soils. Good source of firewood.
Beech, *Fagus sylvatica*	Best on light soils but will stand most soils if the drainage is adequate. Hardy but no good in a stock hedge, for the animals will eat it. A decorative copper-leaved variety is also available.
Buckthorn, *Hippophae rhamnoides*	Fast-growing, spiny shrub capable of withstanding sea gales and suitable for sand dunes. It produces many suckers however, and its invasive growth may be too rampant for a normal hedge.
Dogwood, *Cornus sanguinea*	Makes a dense hedge. Suitable for chalky soils, but requires regular pruning to keep it in check. There are several decorative and variegated varieties.
Ramanas rose, *Rosa rugosa*	Attractive, dense growth for a decorative hedge. Goats will eat it. Produces hips that provide food for birds, as well as being a source of rosehip syrup. Requires pruning to shape it when first planted, but once established, only needs light pruning. Can be used as a gap-filler in a mixed hedge. *Rosa stylosa* is also available.
Wayfaring tree, *Viburnum lantana*	Common tree in existing hedges. Fairly weak growth, but a good source of berries for birds. Does well in chalky soils.
Common alder, *Alnus glutinosa*	Fast-growing, hardy trees suitable for growing as windbreaks in wet soils. Often used where there is a hedge in association with a ditch. There are a number of different types, such as the Italian and grey alders.
Guelder rose, *Viburnum opulus*	Good 'conservation' and gap-filling shrub, providing food berries for the birds and colourful autumn foliage.
Willows, *Salix* species	Excellent in very damp situations where a shelterbelt or windbreak is required. Hybrid varieties that are vigorous and quick-growing are the best choice. Some suitable as biomass fuel plants, others for making baskets or cricket bats! Not suitable for stock control. Goats relish the branches and it is often grown as a source of browsing.

hedge trimmer. A newly planted hedge may only require to have the leaders – the top, primary twigs – cut back in order to stimulate the laterals, or side growth; but be guided by the suppliers' recommendations in relation to specific species. However, newly planted evergreens should not have the leaders cut until the appropriate height is reached.

If a hedge is really overgrown, with tall, lanky growth and gaps at the bottom, it may be best to lay it. Once this is done, it will not need further attention for a number of years, except for an annual trim. Laying involves cutting part of the way through the main stems of the shrubs with a sharp billhook, and then bending them over to one side at an angle of around 30 degrees. They are held in place by stakes that are hammered in so that the pleachers can be intertwined through them. If the ground is on a slope, the stems should always be laid uphill so that they are less likely to break right through. Separate pliable hazel or willow wands can be used to give added support to the laid hedge, by weaving them in and out at the top. These are usually referred to as binders, and give the hedge a neatly finished look.

There are several variations on hedge laying, with different areas often having their own 'designs'. There is also a certain mystique associated with the craft, implying that much lore and traditional practice must be imbibed before one becomes an expert. We found the whole process straightforward, and our hedges responded well after our initially unpractised hands had finished with them.

Ditches

Ditches are there for a good reason. Although they may be marking a boundary between two areas of land of different ownership, the original reason was to drain the land, and so they need to be kept clear otherwise they gradually silt up, which will cause flooding. Ditching is a back-breaking task to undertake by hand: it is essentially a question of digging up the silt, leaf mould and other dross that has accumulated, whilst ensuring that the original line of the boundary is maintained. If there is a considerable expanse to maintain, it is worth getting a contractor to do it mechanically. Alternatively, it may be possible to hire the necessary equipment.

Ditches are often found in conjunction with banks that, in turn, may have hedgerow trees growing in them. Again, it is important not to undercut the bank in case the root systems of the trees are damaged. The soil, silt and leaf mould that is produced when a ditch is cleared is fertile humus: thrown up onto a bank where trees are growing, it helps to increase the fertility of the soil there. If the ditch is a site-line between two properties, it is obviously doubly important to ensure that the original line is maintained. It is also courteous to inform a neighbour when a ditching or hedge-cutting operation on a shared boundary is to take place.

Stone Walls and Banks

In areas where natural stone is readily available, drystone walls make superb barriers; but they need periodic maintenance. It is not a job for the unskilled, but courses, books and videos are available for training purposes. Although it may look easy, it should be remembered that in the past the walls have been constructed without mortar or cement to hold the stones together. The skill is in selecting suitable stones whose surface irregularities interlock with those of their neighbours. These days, with the ready availability of cement and mortar, stones may be reinforced with this, as long as it is not in a conservation area where they may be required to be repaired in the traditional way.

Walls are constructed by first making a foundation trench about 15cm (6in) deep and 15cm (6in) wider than the wall. This is packed with hard, fist-size stones, and then a double line of string is stretched along the trench, showing where the stones are to be laid. A double row of large, flat stones is then placed along the lines indicated by the string, while the central space is packed with small rough stones, called *hearting*. A central *through-stone* running through the middle from one side to the other, makes the structure more stable. To provide even greater stability, the two sides of the wall usually taper in towards the top. Once the appropriate height is reached, a capstone or *copestone* is placed on the top. This is only one of many methods of drystone wall building. In the past, each area had its own distinctive pattern of construction, and the particular wall types are sometimes named after the areas in which they are found.

In some areas, such as the Snowdonia area of Wales, large sections of slate placed upright and joined to each other with twisted cables are used as boundaries. This is making use of natural resources, particularly where old slate-mining operations have left large areas of slate rubble behind. This type of boundary is suitable for sheep but not for animals such as goats that are likely to try and scramble over, and possibly hurt themselves.

In some places, banks are frequently found. They are made of stone and turf, often with hedging plants growing on top of them. Existing ones can be

Upright sections of slate used to make a fence in Wales.

repaired with stones and turf, or even with welded mesh. If a new section of bank is to be made, erect a double layer of welded mesh and fill the space with rubble and soil. Slant the sides inwards, and wire them together to prevent sagging outwards. Lay turfs on the outsides, pegging them into position. Saplings can be planted in the top as required. It does not take long for the whole structure to green over.

Fences

There are several types of fence, but all require sturdy upright posts placed at regular intervals, normally about 3m (10ft) apart, but closer if the land is on a slope or otherwise uneven. Posts are available in wood, concrete, plastic, fibreglass or angle iron, depending on requirements. Angle iron, for example, is more appropriate for light fencing such as poultry or rabbit netting. Plastic and fibreglass are also better for light use, or as garden fencing where different colours may be appreciated. Wooden posts may be completely circular, or half round so that there is one flat surface, or they may be squared, as is often the case with post and rail fences. These are commonly used on a field scale. Pre-cast reinforced concrete posts are often used to support garden fences, or as corner posts elsewhere.

Whatever type of post is used, it needs to be well braced at the corners or where there is a turn in the fence. The upright post has a strut post wedged against it, at an angle of around 30 degrees. A corner post will require two strut posts, each one parallel with the fence direction. Each strut post is then braced against a plate or stone set into the ground.

Posts need to be set in soils that are adequately drained otherwise they will rot more quickly or become dislodged. In loose soils, they are best concreted into place. Corner posts and gateposts need to be set in deeply; intermediate posts can be at a shallower depth, depending on the type of soil.

Digging the holes is relatively easy with a hand auger. This was one of the best tools that we ever bought: it is essentially a T-shaped steel corkscrew, the pointed end of which is placed on the soil, and then the whole thing is rotated by means of the handles. Some tractors have auger attachments, but this is normally for large-scale use.

Loose soil is taken out with a narrow spade, for the narrower the hole the better. Once it is deep enough, the post is inserted. A useful tool for ramming it into place is a post driver. This is essentially a heavy, metal tube, capped at one end, that fits over the post. It has a handle on either side so that two people can ram it down. Depending on the soil, the post then has stones or concrete placed around it, followed by soil that is tamped down.

Post and Rail

Post and rail fences have horizontal rails nailed into the uprights; the height and number of rails will vary, depending on the livestock. Some animals may push their way through gaps in a post and rail fence, in which case covering it with netting is a good idea. Pigs are particularly good at getting their snouts under a barrier, so it needs to be well dug in, or pegged down. A good pig fence can be provided by nailing on galvanized sheets to a post and rail fence; alternatively, pig netting can be used.

Post and Wire

Wire can be used instead of rails, and is a cheaper option. It is important to strain it properly so that it is taut enough, otherwise it will sag; however, don't overdo it in case it breaks and whips back at you! Wire strainers are available from agricultural and horticultural suppliers, and it is a good idea to buy or hire one of these if you have a lot of wire fencing to erect.

Post and Netting

Wire netting can be used instead of rails, or in addition to them. Again, it is important to brace it properly, either by running a wire through the top so that it does not sag, or by stapling it to the cross-rails if these are available.

There are different types of wire netting, depending on the use for which they are required. Chicken, sheep and pig netting have different gauges and come in differing heights. There is also a range of green plastic-coated wire netting that is more appropriate in a garden setting.

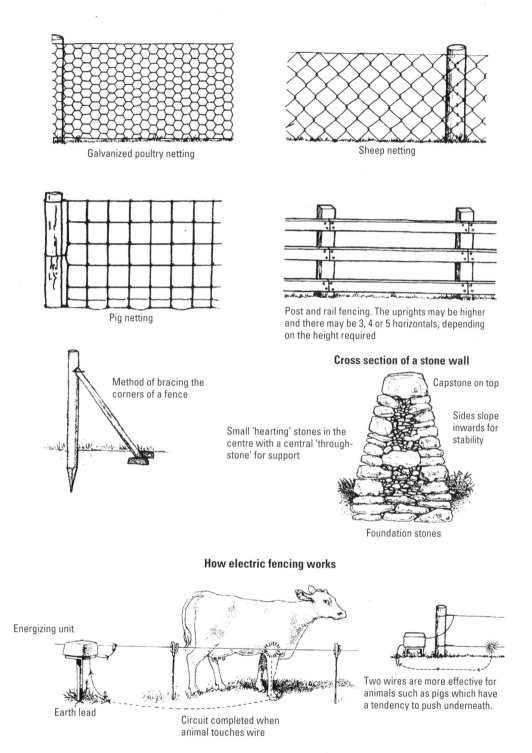

Galvanized poultry netting

Sheep netting

Pig netting

Post and rail fencing. The uprights may be higher and there may be 3, 4 or 5 horizontals, depending on the height required

Method of bracing the corners of a fence

Cross section of a stone wall

Capstone on top

Small 'hearting' stones in the centre with a central 'through-stone' for support

Sides slope inwards for stability

Foundation stones

How electric fencing works

Energizing unit

Earth lead

Circuit completed when animal touches wire

Two wires are more effective for animals such as pigs which have a tendency to push underneath.

Types of fencing.

This post-and-rail fence has had poultry netting added to it. There are also two electrified wires at the top and bottom to keep out foxes.

Pigs do not need a high barrier, but it must prevent them from 'burrowing under'. Here, pig netting is used in conjunction with post-and-rail fencing, and is well dug in to confine this Tamworth.

Chestnut Paling

This type of fence is made up of wooden stakes sharpened at both ends and held together by twisted wire. It is easy to roll up and transport, and can be hammered in quite easily in between stronger supporting posts. The sharpened ends, however, do pose a danger to livestock such as goats that have a tendency to try and scramble over. They can become impaled. If such a fence is already in existence, it can be made safer by covering with wire netting and folding it over the top.

Barbed Wire

Barbed wire is often used as a top strand to stop livestock going over the top of a fence, or as a bottom one to prevent them pushing underneath. Many people dislike using it, however, because of the damage it can cause; many dairy animals have suffered wounds to their udder because of it. We always avoid using it.

Hurdles

Metal sheep hurdles make excellent temporary pens, inside and out, and are also good for temporarily blocking gaps in a hedge. They are available in wood or metal, with the latter being the most common option. They are easily linked, so that any size or shape of pen is possible. They can also be used to form a 'race' or route along which the animals can be driven into the area required.

If you want to have a go at making wattle hurdles, and have a plentiful supply of hazel or willow, it is an excellent rural craft to learn. A baseboard with holes is required, into which upright posts are placed. Split canes of willow or hazel are then woven in and out of the uprights, making a sturdy barrier. The canes need to be supple and may require soaking before use. If a selection of long, thin stems is available, and they are pliable enough, they can be used whole.

Electric Fencing

Electric fencing is the best solution for those with livestock that is particularly hard to confine. It is also useful where controlled grazing is necessary: this allows grass or a fodder crop to be eaten down in one area before the stock is moved on to the next grazing section. For keeping out predators such as the fox, there is nothing to beat electric fencing.

An electric fence circuit has four parts: the controller or energizing unit that provides the electricity; conductor wires held by insulators; the livestock to be controlled; and the ground on which the fence is erected. The controller produces pulses of high voltage electricity, and these travel to the fence wire via the output lead. The earth lead of the controller is securely earthed to the ground. The circuit is incomplete until an animal touches the wire and receives a shock as the electric pulse goes through its body to the ground. The voltage is not enough to hurt the animal, but is sufficient to make it give the fence a wide berth in future.

Energizers are either mains or battery operated. The former must always be installed inside a building, and the instructions from the manufacturer followed precisely. It goes without saying that all mains electrical installations should be carried out

Metal hurdles that join together are useful for making temporary pens. They can also be used to make a 'race' to direct animals to the area where you want them to go.

by a qualified electrician. Mains-operated systems are normally used for permanent perimeter fencing, or for large-scale operations.

Battery-operated energizers are more commonly used on a small scale, or where a fence is moved frequently. They usually operate from rechargeable batteries, either two 6v batteries or one 12v battery. Wind-powered generators to energize the fencing system are also feasible.

If there is an existing perimeter fence or hedge that merely needs supplementing, then one or two strands may be sufficient, erected just in front of the existing barrier.

Where access to the land needs to be controlled on a daily or weekly basis, it is not practicable to move a permanent fence. Lightweight electric netting is available for poultry and livestock and also for keeping out wild rabbits and foxes, but it should not be used for horned animals because of the risk of their becoming entangled. It can be dismantled, rolled up and re-erected further on. The netting is made of polythene and stainless steel conducting twine, and is erected with support poles and ground spikes. Grass must be kept short where an electric fence is erected, otherwise the growth will earth the current and the fence ceases to function.

Windbreaks

The function of a windbreak is not to act as a boundary (although it may do so) but, as the name implies, to provide protection from the wind. This is especially important for young stock, although adults also require shelter in this way.

At its simplest, a windbreak may be an existing hedge or wall, or even a copse of trees. Where a

Traditional hurdles are made by hammering uprights into a baseboard and then interweaving pliable branches or lathes.

Electric poultry netting used to prevent access to a particular area of grass, as well as keeping out foxes.

Here, horticultural mesh has been used to provide a windbreak for these emus.

windbreak is required in an otherwise open area, or on a temporary basis, it will be necessary to construct one that is easily erected and dismantled. A line of straw bales makes an effective shelter for shorter animals and poultry, and metal hurdles with straw bales to fill the gaps are even stronger. Wattle hurdles are windbreaks in their own right, requiring no extra material other than supports or bracing to prevent them being blown or knocked over.

A series of poles knocked into the ground can be intertwined with brashings or leafy twigs to make a temporary windbreak that is suitable for poultry and waterfowl. A double line of wire netting filled with straw is effective for sheep and pigs.

Tightly stretched, close-woven plastic mesh is ideal for tall, exotic birds such as emus and rheas. The mesh breaks the force of the wind but avoids the downward spiralling that can sometimes occur with tall, solid barriers. It would not be appropriate for large animals however, for they would tend to push against it.

The Small Wood

There are many areas on a smallholding where trees can be grown as a woodlot, in addition to boundary hedges and windbreaks. They can be usefully planted in places that are of no use for any other purpose, for instance damp, poor or low-lying ground, hilly areas, awkward corners of fields where machinery cannot gain access, and field perimeters.

Trees can be planted for many reasons: as an investment crop for felling; for coppicing for a particular purpose (the upper part of the tree is cut down, leaving a stump from which new shoots grow, for subsequent cutting); or for generally improving the landscape, usually referred to as 'amenity' planting. Whatever the motivation, planting trees is a long-term investment that enhances the planter, the locality, and ultimately the planet.

Trees from Seeds and Cuttings

Reference has already been made earlier to the planting of saplings. What is often not realized is that trees are also quite easy to grow from seed, and seed companies offer a wide range of varieties. The seeds can be sown in seed compost, in containers that are taller than normal, to allow for the rapid growth of the root system; once germinated and growing strongly, the containers are put outside in a nursery bed. This could be a layer of sand on which the pots rest, with netting over the whole lot to protect against wild rabbits and deer.

Most trees are hardy, and the saplings will require little more by way of aftercare than regular watering, and possibly shading in the event of particularly hot sun. When they are about 45–60cm (1ft 6in–2ft) high, they are ready for planting in their permanent positions. Some people have found that they can sell their surplus tree seedlings quite easily, particularly to garden centres or to those interested in bonsai cultivation.

Trees for Felling

Some trees, such as South American beeches, are

Livestock Fencing Requirements

Livestock	Suitable fencing	Height	Comments
Chickens	Poultry netting. Stone wall. Thick hedge. Electric fencing/netting	1.5m (5ft)	To exclude fox, 2m (6ft 6in) required unless electric fencing is used
Ducks	Thick hedge. Stone wall. Poultry netting	60cm (2ft)	Generally easy to confine. Exceptions are Muscovy ducks. Fox must be excluded, as above
Geese	Thick hedge. Stone wall. Poultry netting	90cm (3ft)	Must be protected against foxes
Pigs	Stone walls. Post and pig netting. Electric fencing	90cm (3ft)	Must be really strong and well anchored, otherwise they use their noses to push under
Sheep	Hedges. Stone walls. Banks. Post and rail fence. Hurdles. Electric fencing	90cm (3ft)	Higher for primitive breeds because of tendency to jump
Goats	Stone walls. Post and rail fencing. Electric fencing	At least 1.5m (5ft) with no large gaps	May eat through hedges. Milking goats are easier to confine than goatlings, which may climb through gaps or jump over a fence. Electric fencing is effective
Cattle	Thick hedges. Stone walls. Post and rail fencing. Electric fencing for controlled grazing	1.4m (4ft 6in)	Bullocks tend to lean against fencing and if the posts are weak, may knock it down. Electric fencing is effective
Horses	Hedges. Walls. Post and rail fencing	1m (3ft 3in)	If you have show jumpers, you may need higher fences
Alpacas	Post and rail fencing. Sheep netting	90cm (3ft)	Avoid any barbed wire
Llamas	Post and rail or netting	1.2m (4.5ft)	Llamas will chase away foxes
Ostriches	Strong, high-tensile wire fencing with 'spacers'	1.8m (6ft)	Warning sign on the fence required. Windbreaks also necessary
Emus	Post and high tensile wire as above	1.8m (6ft)	Warning signs on fence needed. Windbreaks needed
Rheas	Post and rail or wire	1.5m (5ft)	More docile than other ratites

grown for timber production and then felled when they are big enough. Poplar is also a popular choice of those who wish to plant trees as a long-term investment crop, although modern hybrid poplars are extremely fast-growing. In this case, the wood is primarily used by the building and carpentry trades. The cricket-bat willow, *Salix alba coerulea*, is another worthwhile investment crop where the ground is damp and unsuitable for other agricultural uses. In Britain, it is possible to buy young saplings from specialist suppliers who will then buy back the timber crop when it is felled and use it for the manufacture of cricket bats. There are also fast-growing hybrid willows that root rapidly from cuttings put in a trench to three-quarters of their length. They, too, leaf very early in spring.

Grants are available from a number of sources for amenity tree planting in the British countryside. Most district councils and county councils have a Forestry Officer who will ensure that those who want help receive sound, professional advice. The Forestry Commission will give grants in respect of areas of tree plantings of 0.25 hectare (1 acre) or more, as long as the aim is timber production. Both conifers and broadleaved trees are eligible, but the broadleaved ones attract a higher rate. Applicants will be required to work to a five-year approved plan.

Coppicing

Coppiced trees are those that are cut leaving a stump so that new shoots are produced; these grow upwards, and are then cut for a particular reason. The willow, *Salix vimnalis*, produces shoots or osiers for basketry, while the fast-growing hybrid willows, referred to above, are frequently grown as an energy crop. Hazel, *Corylus avellana*, has coppicings used in the manufacture of hurdles, and sweet chestnut, *Castanea sativa*, is used for making fences. Where young trees are planted in order to establish a coppiced area, they should be spaced 2.4–2.75m (8–9ft) apart. Other details of planting are the same as those detailed earlier.

Coppicing is carried out in winter, so that the whole of the following season is left for new growth. The cut should be as close to the ground as possible so that the new shoots grow from the rootstock rather than the cut stem, and it should slope to one side so that water can run off.

Christmas Trees

The species normally used is the Norway spruce, *Picea abies*, but others for which there is a demand include blue spruce, *Picea pungens glauca*, Scots pine, *Pinus sylvestris*, and lodgepole pine, *Pinus contorta* var. *latifolia*. Firs are also used, including Noble fir, *Abies procera*, Nordmann fir, *Abies nordmanniana*, and Douglas fir, *Pseudotsuga menziesii*.

The ideal site is moist and free draining, where protection can be provided against rabbits, hares and deer. Christmas trees can be planted between other trees, but they are more commonly planted in their own plantations, where the usual spacing is 1sq m (11sq ft). Extra feeding is not normally required if some fertilizer was included at the time of planting. If the foliage looks poor, however, a nitrogen foliar feed can be given in the summer before harvesting. It is important to keep down weeds, otherwise the foliage of the bottom branches of the trees may become brown.

Christmas trees should not be lifted or cut before they are dormant; in the northern hemisphere, this is generally the latter half of November. Anti-transpirant sprays can be applied in order to reduce needle drop in Norway spruce; this can be carried out just before lifting.

When the last trees have been cleared, a year of fallow is recommended, in order to allow time for the destruction of perennial weeds, or to increase the humus content by green cropping. This will also be a time to check the soil fertility before the next rotation is commenced.

Nut Trees

The most common nut trees are filberts (hazels) that are often grown as hedge plants, and particularly as internal screening for commercial orchards. For a good crop of nuts it is necessary to plant different varieties such as Kentish cob, Cosford cob and purple-leaved filbert, but including some wild hazels will ensure that all hybrid cultivars are pollinated. Cosford cob is a cultivar that will also pollinate other filberts.

Walnut trees are normally planted as 'stand-alone' trees. Walnuts should not be planted near apple trees because the root exudations have a restrictive effect on the latter. A good variety that can fruit within four years if it is in compact form, is Broadview. It is also self-fertile.

Wood-Burning

Where there is existing woodland on a farm, it may be worth looking into the possibilities of coppicing some of the growth and selling it as fuel to local customers with wood burners, or to harvesting it for your own use. (*See* the wood-burning section in Part I.) The hybrid willows referred to above are particularly suitable for this, although willow was not traditionally regarded as a good source of firewood, as its omission from the rhyme shows:

> Beechwood fires are bright and clear if the logs are kept a year.
> Chestnut's only good they say, if for long 'tis laid away.
> Birch and fir logs burn too fast, blaze up bright and do not last.
> It is by the Irish said, hawthorn bakes the sweetest bread.
> Elm wood burns like churchyard mould. E'en the very flames are cold.
> Poplar gives a bitter smoke, fills your eyes and makes you choke.
> Apple wood will scent your room with an incense like perfume.
> Oak and maple, if dry and old, keep away the winter's cold.
> But ash wood wet or ash wood dry, a king shall warm his slippers by. (Anon)

5 The Kitchen Garden

I have banished all worldly care from my garden.
(Hsieh Ling-Yin AD 410)

The kitchen garden is a personal thing. Once regarded as a purely utilitarian concept where only edibles were grown, it is increasingly seen as a place that is attractive to the eye as well as to other senses. Here may be found a range of vegetables, herbs, flowers and companion plants, inter-planted to mutual benefit and good effect. Vegetables take priority, however, and as there is such a wide choice, it is a good idea to make some early decisions about what to grow, and on what scale. Our own plans were based on the following factors:

- crops that everyone in the household will eat;
- crops that can be grown organically without too much difficulty;
- varieties with good flavour and disease resistance;
- crops suitable for the particular soil and conditions;
- crops that produce a reasonable yield in relation to the ground occupied;
- crops that cannot be bought cheaply locally.

Growing crops without using chemicals is not without its problems, but much can be done to avoid them. As a rule, early varieties are more free of virus and insect pests than later ones, although they are more at risk from hungry birds. Netting can be used as protection in this case.

It makes sense, as far as possible, to grow those varieties that are disease resistant. When a variety is so described, it does not mean that it is absolutely safe against a particular disease, merely that it has a greater resistance to it than other varieties.

On the question of varieties, flavour is an important consideration. This may mean that some commercial varieties are discarded on the grounds that, although the yield is high and the produce of a uniform size, the flavour is poor. Choice of varieties is, of course, a personal one, and it is essential to study the seed catalogues every winter to see what is available.

It is sensible to concentrate on varieties that are suitable for a particular soil. Yield of crop in relation to ground used is also important. Some crops such as runner beans that grow upwards produce a heavy yield and yet occupy a comparatively small area of

Part of the author's kitchen garden.

ground. Brussels sprouts produce a lower yield in relation to ground occupied, but it is one of the best winter vegetables when there is little else available. It may not be appropriate to grow some varieties such as main crop potatoes if they are available relatively cheaply from local growers.

There is an unending supply of gardening tools, but the basic ones are a spade, a fork and a hoe; small hand versions of these are also useful. With these, most gardening tasks can be undertaken. Mattock-type tools are widely used for cultivation in Europe, and many people claim that they are much easier on the back. It is worth paying extra for good quality tools. Stainless steel keeps its sharpness, and heavy, sticky soil is less likely to adhere to it. Whatever the choice, tools will last much longer if they are cleaned regularly and stored in a locked, dry shed. Rubbing with an oily rag after each usage will maintain the tools for a lifetime.

Soil Preparation and Maintenance

No crops will thrive unless they are given good quality, fertile soil, and the optimum conditions for their individual needs. The first step is to prepare the soil and then have a clear plan of how it should be used and maintained. Taking an area of grassland or other vegetation as our starting point, the main steps are as follows.

Clearing the Ground

It is rare for anyone to start with a clear area of soil, and in most cases it is necessary to clear the ground of turf or long-established weeds. The hardest way of doing this is to cut out the top growth, square by square, skimming it off with a sharp spade. If it is turf, however, the squares can be stored elsewhere, grass-side down and left to break down into fertile soil.

A practice followed by many organic growers is to burn off the growth with a horticultural flame gun; however, although this clears and sterilizes the topsoil, the underlying roots must still be dealt with. Alternatively, a glyphosate product can be sprayed onto the plant growth. This is taken in by the leaves, sent down to the roots and kills off all perennial and annual weeds. It is inactivated as soon as it touches the soil. It is not an organic product, but is safer than many other weedkillers, and planting can take place as soon as the ground is dug over.

Some organic gardeners prefer to use ammonium sulphamate that kills off most weeds but not nettles and thistles. It is non-selective, so the ground must be left for six months before planting, and the nettles

and thistles must still be cleared. Meanwhile, the product gradually decays into ammonium sulphate that acts as a fertilizer. If there is no rush to bring an area of land into cultivation, the weeds can be smothered over a period of time by excluding light: black plastic or sections of old carpet are effective in killing off surface growth, leaving the area ready for digging after about twelve months. This practice also encourages earthworms that help to aerate the soil.

Housing a pig or some chickens on the area to be cultivated is an option. Their indiscriminate rooting or scratching activities help to get rid of weeds, roots and insect pests. Unfortunately, earthworms and beneficial insects are taken as well. It is important to allow the livestock access to a small area at a time, so that it is completely cleared before they are moved on.

Cultivation

Once the top growth and perennial weed roots have been removed, either by hand or chemically, the ground can be brought into cultivation. There are two ways of doing this: digging by hand, or using a mechanical cultivator.

Digging by hand is a slow business, but there are many who claim that the resulting soil is better because any remaining roots can be removed. As digging proceeds, rotted manure or compost can be added. How this is done is largely a matter of preference: it can be spread over the whole surface first so that it is incorporated as you go, or it can be forked from a wheelbarrow into the particular trench on which you are working.

Using a mechanical cultivator is much quicker, but it is important to remove all perennial weed roots first, otherwise they will be chopped up and distributed all over the bed, and each piece will then grow into a new weed plant. It is also important to ensure that manure or compost is completely rotted, otherwise nitrogen depletion will occur. Nitrogen can be stolen from the soil as the decomposition process of the manure continues.

Once the soil has been turned, the weather will help to break down any remaining clods. Before planting or sowing, however, the soil should be well raked to bring it into a fine crumb consistency.

Liming

The level of acidity or alkalinity of a soil is referred to as its pH (potential hydrogen) value. This can be measured by taking soil samples and conducting a simple test on them. Soil-testing kits are readily available in garden centres, while farm supply

Preparing the ground

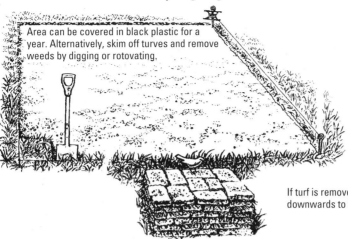

Area can be covered in black plastic for a year. Alternatively, skim off turves and remove weeds by digging or rotovating.

If turf is removed, stack it grass side downwards to form loam for future compost

Making a raised or deep bed

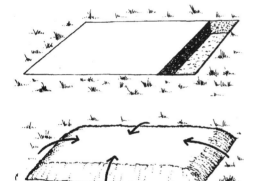

Dig trench and break up sub-soil. Then throw top soil from second trench into first, incorporating generous amounts of well-rotted manure and compost. Continue until completed.

Rake the soil from the sides inwards. If necessary place supporting planks along the sides to retain the soil. Do not walk on the soil to compact it.

Comparison of normal and deep bed

Normal bed

Raised beds were common in medieval times

In a deep bed there is more top soil, allowing roots to grow deeper, with less lateral competition. The plants can therefore be grown closer together

Raised and deep beds.

agencies normally stock larger kits suitable for a wider range of tests. Alternatively, there are companies whose services include coming on site to do the testing for the grower.

Most plants will grow well where the pH is around 6.5, although there are obvious variations. Brassicas, for example, are at their best on a more alkaline soil of pH 7.0–7.5, while heathers and azaleas favour a more acid soil.

Soils vary tremendously, from heavy clay to light sand, and from marshy areas to quick-draining limestone soils. All types of soil benefit from the addition of compost, but those that are heavy or tending towards the acidic will need to be limed. Where the soil is a clay one, the lime has a flocculating effect in that it helps to break down solid lumps of clay, redistributes individual clay particles and improves the structure of the soil. The effect of frost on roughly dug heavy soil will also help to break down the clods. Lime is also needed to unlock other elements in the soil that might otherwise remain unavailable to the plants. It counteracts acidity and has a general 'sweetening' effect on the soil, as well as making the earth more attractive to the earthworm population. It is important not to apply it at the same time as manure otherwise there will be a reaction between the two and the fertility will be decreased. If the initial digging is done in the winter, the lime can be added at this time, while the manure is left until the spring, or vice versa if preferred.

Soil Fertility

Unless bought-in fertilizers are being used, it will be necessary to maintain the fertility of the soil in other ways. Crops will use up the available nutrients, and these must be replenished if reduced cropping and damage to the soil structure is to be avoided. The fertility can be maintained in several ways.

Composting

Where livestock are kept, there will be a plentiful source of organic matter, but it needs to be stacked and allowed to rot down completely before being added to the soil. A convenient way of doing this is to use compost containers that will not only keep the stack in a tidy condition, but will also ensure adequate aeration and warmth to accelerate the breaking-down process. Compost containers can be made from timber or bought as complete units. To make a DIY one, wooden posts can be sunk into the ground to make the four corners of a box. Three sides are then made by nailing wooden boards

across, leaving small gaps for aeration. If battens are nailed to the two uprights on the fourth side, it allows a gate structure to be installed that slides up and down, thereby opening and closing the heap. A bank of such containers ensures that compost is available on a regular basis. In cold areas, decomposition is accelerated by insulating the inside of the container with pieces of old carpet and placing a piece on top, weighted down with some bricks.

Used bedding straw, poultry litter, grass cuttings and annual weeds form a good basis for compost heaps. An extra layer of animal manure, such as rabbit or chicken droppings, every few inches heats up the heap and hastens the decomposition process. Alternatively, it is possible to purchase compost acceleration products. Adding a sprinkling of lime helps to encourage earthworms that, in turn, accelerate decomposition.

The roots of perennial weeds such as docks and nettles, as well as twigs and other material that would take a long time to rot down, are best avoided. Inevitably a certain proportion of such things find their way in, and, it must be admitted, our compost has been known to include pieces of plastic and baler twine as well as the long-lost potato peeler. Although potato and other vegetable peelings are worth adding to the heap, it is not a good idea to include meat or other food leavings, otherwise rats are encouraged.

Tree leaves produce excellent compost, but as they take longer to rot down to leaf mould because of their lignin content, they are best stored on their own: four corner posts with wire netting walls makes a suitable container. Ours was always popular with hedgehogs.

On a small scale, garden refuse can be put into black plastic bags, sealed and left to decompose in an out-of-the-way spot until required. Some of the best compost that we ever produced was the result of this anaerobic (without air) decomposition. It was a pleasant surprise to come across the bags that had lain forgotten behind the garden shed.

Compost ready in late autumn and winter can be incorporated during winter digging. If the compost is not completely rotted at this time, it can be merely laid on the surface, rather than being dug in. In this way, nitrogen loss from the soil is avoided, and weeds are prevented from becoming established; it encourages earthworms to take up residence underneath and they, in turn, help to break it down. By the time spring comes along, it is ready to be dug in.

Fertilizers

Home-made compost and rotted manure provide bulk and natural fertilizers in the soil. It is also

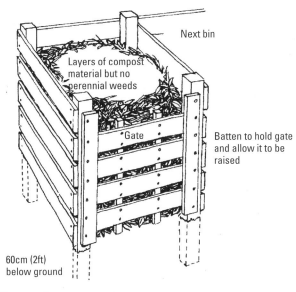

Next bin

Layers of compost material but no perennial weeds

Gate

Batten to hold gate and allow it to be raised

60cm (2ft) below ground

Constructing a compost heap.

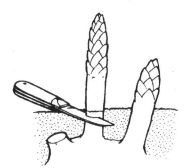

Cutting asparagus spears.

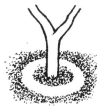

Circle of lime to guard against clubroot

Tarred paper collar to deter cabbage root fly

Protecting brassicas.

Make a hole with a dibber and drop the leek plant in. Do not fill the hole with soil but water in

Trim the leaves and roots before planting. Have only the green part above soil level

Planting leeks.

possible to buy in organic and inorganic fertilizers for adding as necessary; these may be dry or liquid. It is important to follow the manufacturer's instructions when it comes to dilution and application. A dry fertilizer can be applied as a 'top dressing' to the soil surface, while a liquid one can be diluted and used as a foliar feed where it is sprayed onto the leaves of growing plants. Some dry fertilizers are available as granules and are described as 'slow release' because they feed the soil gradually over a period of months. They are generally more expensive than other fertilizers.

The most important elements required by plants are nitrogen (N), phosphorus (P) and potassium (K). Nitrogen caters for leaf growth, and a deficiency is indicated by small, yellowing leaves. Phosphorus provides for healthy root growth and fruit production, and a shortage is indicated by inadequate growth and dark or reddish leaves. Potassium is needed for flowers and fruit, as well as good disease resistance, and a deficiency shows up as weak growth and susceptibility to disease.

A range of compost containers that can be used in sequence.

Comfrey can be used to make a natural fertilizer for most plants, but it needs to be grown in its own bed, for it can be invasive.

Bought fertilizers indicate the NPK balance on the label. One that contains all three in the same amounts is a balanced or general fertilizer. Special fertilizers will have a different ratio depending on the specific use for particular plant groups. Many fertilizers will also contain trace elements such as boron and magnesium that are needed in tiny quantities. Some bought foliar feeds contain trace elements and are used to give a boost where deficiencies are suspected.

It is possible to make a good quality liquid fertilizer by using horse, rabbit, chicken or sheep manure, placed in a bag and suspended in water, in a butt. Alternatively (or in addition to), comfrey and nettle leaves can be used in the same way, adding more to the butt as they break down. The resulting product smells awful and it is essential to keep the butt covered. If it has a tap, it is easy to extract the liquid as required. In both cases, dilute it in the ratio of 1 part fertilizer to 16 parts water until it resembles weak tea; it is then ready to be applied.

Green Manuring

Green manuring is the practice of sowing an annual cover crop on a fallow or vacant bed so as to prevent weed seedlings from becoming established. The plants protect the bed from wind erosion as well as weeds, and prevent soil compaction taking place. Green manuring is frequently carried out in late summer to autumn after an earlier crop has been harvested. When the plants have grown they are then dug in when winter or spring digging is carried out, adding fertility (green manure) to the soil.

Some cover-crop plants will fix atmospheric nitrogen in their root nodules, adding to the nitrogenous content of the soil. They include lucerne (alfalfa), broad beans, red clover, annual lupin and winter tares. Other green manure plants are Italian ryegrass, rye, buckwheat, mustard and phacelia.

Planting

When we had carried out the initial ground preparation for our kitchen garden, we divided it into several small plots, rather than having one large area. It is easier to work a plot if there is access from both sides, without treading and compacting the soil. There is also a sense of achievement at completing each small bed, compared with the dismay when confronted with one large one.

Raised or deep beds are appropriate for a bed system: these are essentially beds cultivated to a greater depth than normal. The increased topsoil allows crops to be grown closer together because there is less lateral root competition, and the leaf canopy of the plants themselves provides a 'living mulch' to resist drought and suppress weeds. The medieval gardeners of Europe grew crops in this way, as well as later market gardeners. What caused its decline was the changeover from horse-drawn transport to the motorcar, with a resultant shortage of horse manure. The same process of mechanization brought about a situation where crops were increasingly grown on a field scale, in straight lines, rather than in the traditional 'staggered' planting, which was more labour-intensive. The 'staggered' planting meant that there was less space wasted and a greater yield was possible, but it needed smaller beds that would not have their soil compacted by being walked on.

A range of protective devices in the vegetable garden, including wire netting, plastic netting and plastic bottles with the bottoms cut off.

The beds are raised about 15cm (6in) above the path level, with wooden boards used as boundaries. Some beds can be left without boards so that the sloping sides can be used for growing short-rooted crops such as lettuce, radish, alpine strawberries and nasturtiums. A width of 1.2m (4ft) makes it easy to gain access from either side of the bed, while the length will be determined by the amount of space available.

Crop Rotation

A crop rotation system is where crops are grown on a new area of ground each year so that the risk of soil-borne diseases is lessened. Inter-planting in a traditional French 'potager' style is still possible, but it makes sense to keep roots, legumes (peas and beans) and brassicas (cabbage family) apart. A potage garden is where different crops and decorative flowers are grown together, producing a visually attractive and functional kitchen garden.

The practice of monoculture, where one crop is grown to the exclusion of others, is a comparatively recent idea, and many authorities believe that it, more than any other factor, has resulted in an escalation in the problem of plant pests and diseases. We always found a four-year crop rotation system to be satisfactory (*see* panel).

Companion Planting

There is a long, mainly oral tradition of growing certain plants together. The logic is that some plants have a mutually beneficial effect, while others are antagonistic and therefore have a damaging effect. The problem is that there are too many variable factors to test this theory on a scientific basis. Also much of the horticultural research taking place is linked with commercial growing, which relies heavily on chemical products, and the chemical companies have no reason to research a subject which would be against their own financial interests. So companion planting is still in the realm of old wives' tales, but this is not to say that it should be dismissed as such. Old wives' tales usually contain a grain of truth; the problem lies in finding it amongst all the accumulated dross. Like many gardeners, we have adopted companion planting on the basis that it may work and is therefore worth trying.

Reading about the theory of companion planting is one thing, but putting it into practice is another. A crop rotation system can easily be swamped by a mishmash of different plants, to the extent that the rotation sequence becomes confused. Dwarf beans and cabbages are said to be good companions, but if they are inter-planted, what becomes of the crop rotation system which has the former in a legume bed and the latter in a brassica bed? Some of the advice written about companion planting is not based on common sense or experience. Peppermint planted with the cabbages may give protection against the cabbage white butterfly, for example, but history does not relate how you are supposed to clear the peppermint out at the end of the season.

A Four-Year Crop Rotation

First Year

Bed A	Bed B	Bed C	Bed D
Potatoes	*Legumes* (Peas and beans)	*Brassicas* (All members of Cabbage family)	*Roots* (Beetroot, carrots, etc)
Prepare ground in winter incorporating rotted manure. No lime.	Prepare ground in winter incorporating rotted manure. Add lime in spring before planting. Leave plants to die down naturally.	Prepare ground in winter incorporating rotted manure. Add lime in spring and sprinkle around plants after planting.	Prepare ground in winter. No manure. Add lime in spring before planting.

Second Year

Legumes Prepare ground in winter incorporating rotted manure. Add lime in the spring before planting.	*Brassicas* Remove top growth of legumes only. Prepare soil in winter adding rotted manure. Lime in the spring.	*Roots* Prepare ground in winter. No manure. Lime in the spring.	*Potatoes* Prepare ground in winter incorporating rotted manure. No lime.

Third Year

Brassicas	*Roots*	*Potatoes*	*Legumes*

Fourth Year

Roots	*Potatoes*	*Legumes*	*Brassicas*

Ground preparation for years 3 and 4 is as indicated for years 1 and 2. The fifth year is the same as year 1.

Faced with realities such as these, we made various compromises and decisions based on a more practical approach. Thus, the four-year crop rotation system would be maintained. Any companion plants used, within the rotation system, would be annuals, so that bed clearance was simplified. Perennials would be in permanent positions within perennial beds that were outside the rotation system. No weeds would be in or near the beds. Yarrow, for example, is said to enhance essential oil production, and some people recommend planting it along borders or with herbs. We grow it mainly for cutting and adding to the compost heap to provide phosphorus and other minerals, but its growth is so rampant that it needs to be kept in solitary confinement.

Weeding

Weeding is a task that every gardener must face. Reference has already been made to the necessity of removing perennial weed roots before cultivation of the soil can begin. Some common perennial weeds include nettles, thistles, bracken, bindweed, mare's tail, yarrow, couch grass, plantains, docks, dandelions, buttercups, hogweed and ground elder.

Once seeds have been sown or seedlings planted out, the main task will be to cope with annual weeds that sprout as if from nowhere. They are much easier to deal with as long as they are not allowed to seed. The most common annual weeds are shepherd's purse, groundsel, chickweed, charlock, pimpernel and speedwell.

Chemical gardeners have an armoury of weedkillers at their disposal, but most people would hesitate before using them where food crops are concerned. There is no organic method of dealing with annual weeds, other than to use a good hoe to decapitate them. They will then die off on the surface. Regular hoeing is essential throughout the growing season and certainly before the weeds seed themselves. Depending on the system of planting, a full-sized hoe or a small hand one can be used.

Squash plants require a lot of water.

Regular hoeing keeps down the annual weeds.

Mulches are useful in keeping down annual weeds. A mulch is a layer placed on the soil surface in order to conserve moisture and suppress weed growth. There are many options including compost, cocoa-shell, straw and bark chippings. Peat is a non-renewable resource that environmentalists are trying to protect, but coir (coconut fibre) is an acceptable alternative available from most garden suppliers. Black plastic can be used to good effect, as well as cardboard and newspaper, although both the latter are unsightly. Lawn clippings can be used as long as they are not allowed to touch plant stems, when they will risk scorching them.

Vegetables

Artichokes
Light, fertile soil. pH 6.5
There are several different artichokes. The globe artichoke is grown for its plump heads; with its tall growth and spread it requires a lot of room, as well as wind protection. Rooted suckers from a variety

such as Vert de Laon are planted 5cm (2in) deep and 90cm (3ft) apart in the spring. In the first season, no cuttings should be taken, and the plants should be allowed to die down naturally. In cold areas, winter protection such as straw may be needed. In the second year the heads can be cut as soon as they are well filled; at this stage they are soft before the scales open.

Jerusalem artichokes are tall plants of the sunflower family and grown for their tubers. A variety such as Fuseau has smoother tubers than the knobbly ones found in older varieties. Tubers are usually planted in early spring, 30cm (12in) apart and at a depth of around 15cm (6in) for added stability. As they can grow up to a height of 3m (10ft), they are often used to screen off an area such as a kitchen garden, or to provide shelter for pheasants. In summer the flower heads should be cut off to encourage tuber growth; these can then be dug up from autumn onwards and throughout the winter. It can be an invasive plant, so needs to be kept out of a crop rotation system.

Plants with their Claimed Companions and Antagonists

Plant	Companions	Antagonists
Apple	Tansy, mint, climbing nasturtium	
Artichokes	Parsley	Garlic
Asparagus	Parsley, basil, tomatoes, marigolds	
Aubergine	Dwarf beans, runner beans	
Beetroot	Dwarf beans, onions, kohlrabi	Runner beans
Broad beans	Carrots, summer savory, potatoes, cabbage, petunias, cauliflower, borage, thyme, buckwheat	Onions, chives, garlic
Brassicas	Peppermint, potatoes, dill, sage, caraway, chamomile, broad beans, thyme, hyssop	Tomatoes, runner beans, strawberries
Carrots	Onions, chives, garlic, peas, sage, lettuce, tomatoes, leeks	
Celery	Tomatoes, dwarf beans, brassicas, leeks, lettuce	
Courgette	Nasturtium, borage, fennel	Rue, potatoes
Cucumber	Borage, peas, sunflowers, dwarf beans, lovage	Thyme, sage, potatoes
Dwarf beans	Summer savory, celery, potatoes, cucumber, sweetcorn, petunias	Onions, chives, garlic
Fennel	Courgettes, marrows, squash	Most other plants
Garlic	Beetroot, tomatoes, strawberries, lettuce, roses, raspberries	Beans, peas
Leeks	Carrots, celery	Peas, beans
Marrow	Sweetcorn, nasturtium, borage	Potatoes
Onions	Carrots	Beans
Parsley	Angelica, artichokes, potatoes, asparagus, tomatoes	Lavender
Parsnips	Onions	
Peas	Mint, carrots, turnips, radishes, sweetcorn	Garlic, horseradish, potatoes
Pears	Tansy, mint, climbing nasturtium	
Potatoes	Horse-radish, mint, brassicas, marigolds, pumpkins	Tomatoes, sunflowers
Pumpkin	Sweetcorn, nasturtium, borage	Potatoes
Radish	Nasturtium, peas, cucumber, lettuce, catnip, chervil, dianthus, lettuce, peas	
Raspberry	Rue, garlic	
Runner beans	Summer savory, sweetcorn, petunias	Onion family, sunflowers
Rhubarb	Spinach, parsley	
Soya beans	Strawberries, dwarf beans	Onions, garlic, chives
Spinach	Strawberries, dwarf beans	Cabbage family
Squash	Nasturtiums, sweetcorn, borage	Potatoes, thyme
Strawberries	Spinach, borage, lettuce, dwarf beans	Cabbage family
Swedes	Peas, catnip	
Tomatoes	Parsley, basil, chives, nasturtium, marigolds	Potatoes, brassicas, tobacco
Turnips	Peas, catnip, thyme, radish	

Mulching the crops to keep down weeds and retain soil moisture.

Chinese artichokes produce small tubers from the *Stachys affinis* plant that grows to around 45cm (18in) tall. Tubers are planted as soon as the soil has warmed up. As the plants grow they can be earthed up, and if kept well watered, will produce tubers ready for harvesting in late summer to autumn.

Asparagus

Open, sunny situation. Well-drained soil. pH 6.3–7.5
This is a luxury perennial that used to be grown on ridges, but recent research has shown that it grows just as well in a normal bed. One-year-old corms or young module plants are the best to plant because they get off to a good start. Plant them in spring, 30cm (12in) apart on well cultivated and fertile soil. Keep the bed well weeded, and resist the temptation to cut any of the stalks in the first season. Cut down the foliage when it turns yellow in the autumn, leaving short stalks, and then mound up the soil over

them by drawing it with a hoe from either side. With modern cultivars it is possible to cut some of the larger spears in the second year. Cut them just below the soil surface, but restrict the cutting to a period to around six weeks in early summer. After that, allow the subsequent spears to produce foliage and follow the same practice of cutting down foliage in the autumn, as in previous years.

Recommended cultivars are the new all-male hybrids such as Franklim and Lucullus.

Beetroot

Sunny position. pH 6.5–7.5
A popular pickling vegetable, this needs to be sown direct where it is to grow. It needs adequate water otherwise it will bolt and go to seed. Ensure that there is no fresh manure in the bed, for this will have the effect of making it fork (form two separate roots). An early variety such as Boltardy has some resistance to bolting and can be sown as soon as the soil has warmed up in the spring. A common mistake is to sow too thickly; sow in individual stations in a staggered row, with three seeds in each station. The seeds are big enough to make handling easy and after germination the strongest of the three is left to grow on while the others are thinned out. Some new varieties, such as Modella, are monogerm cultivars and produce only one seedling per cluster so that thinning is unnecessary.

To harvest, lift the beets and twist off the stems; they 'bleed' if cut. They can be stored in sand in cool conditions, or left in the ground until needed, as long as frost protection is provided.

Broad Beans

Open site. Well cultivated soil. pH 6.0–6.5
Broad beans are easy to grow, but there can be problems of mice eating the seeds of early varieties that are sown in autumn for over-wintering. We eventually found that it was better to delay sowing until spring. Indoor-sown plants that were well grown were then transplanted to their final positions when the soil had warmed up.

Blackfly aphids that affect the growing tips can also be a nuisance. The best defence is to sow as early as possible and then to pinch out the growing tips when the plants have reached their full height. Spraying the plants with diluted seaweed extract also helps to deter them, while providing a food supplement for the plants.

Seeds or plants are spaced 15cm (6in) apart in all directions. Useful varieties are Aquadulce and the excellently flavoured Red Epicure.

Removing the growing tip from broad bean plants helps to protect them from blackfly aphids.

Broccoli

Fertile soil. Sunny or slightly shaded position. pH 7.0–7.5

A useful and hardy vegetable, broccoli seems to withstand the coldest of winters. There are basically two kinds: purple-sprouting, such as Early Purple or Claret, and white-sprouting such as Early White. They can be sown in boxes or a seedbed in late spring for transplanting to their growing site in summer; they are then ready for cutting in late winter. The heads are cut before the flowers open.

They are spaced about 60cm (24in) apart, and may need bird protection. As with all brassicas, there is a danger of clubroot infection, and it is important to lime the soil: we sprinkle an added circle of lime around the stem of each plant; this also helps to deter slugs. Cabbage-root fly can be a problem, and collars of tarred paper placed around the stem to foil

the fly will give protection. These are now available at most garden suppliers, or they can be made from rubber carpet underlay.

Calabrese is a bluish-green form of sprouting broccoli, sometimes called American, Italian or green broccoli. It can be sown in succession from spring onwards. It does not transplant as well as other brassicas, and is best sown direct. Varieties include Early Romanesco and Trixie, which has some resistance to clubroot.

Brussels Sprouts

Firm, fertile soil. Sunny or slightly shaded position. pH 7.0–7.5

Sprouts are sown from late winter onwards, in seed trays, for pricking out and subsequent transplanting. As with all brassicas, we prefer to grow our own from seed to avoid the risk of introducing clubroot disease by buying in plants that may carry it. Soils that have adequate supplies of lime are less likely to be affected, so the application of lime should never be neglected.

The plants are put in their final places as soon as they have been hardened off, and are planted at a distance of 60cm (2ft) from each other. Earthing up the plant stems helps to provide greater stability against the winter winds. Supports may be necessary in exposed areas.

They are ready for harvesting as soon as the sprouts are nicely formed, but the flavour is improved after they have been touched by frost, so we always wait for this before picking. Pick from the bottom upwards, and remove any yellowing leaves at the base.

There are early varieties such as Peer Gynt, as well as later varieties such as Sheriff.

Cabbage

Firm, fertile soil. Sunny or slightly shaded position. pH 7.0–7.5

Spring, summer and winter cabbages need similar conditions to other brassicas, and are planted with 15–20cm (6–9in) distance between them, depending on the variety and its size. If all three types are grown, cabbages will be available throughout the year.

Bird attack can be a nuisance, and netting is the best insurance against this. Cabbage white butterflies relish the leaves as a nursery for their caterpillars, and where chemicals are not used, picking off the caterpillars or spraying them with salt solution is necessary. On a larger scale, the plants can be sprayed with a biological control (*see* page 231).

Brussels sprouts are one of the most valuable winter crops. Photo: Elsoms

Chinese cabbage, such as Jade Pagoda, differs from most other cabbages in being comparatively shallow-rooted, so needs frequent watering. It should be sown in mid-summer, and is useful to follow earlier harvested crops such as early peas. There are also loose-leaved Chinese cabbage such as Pak Choi that can be sown in summer for later cropping.

Carrots

Sunny, sheltered position. Light, well drained soil. pH 6.5–7.5

Our soil is rather heavy for carrots. Initially we grew only early varieties that were not deep rooted, but after going over to deep beds, we were able to grow main crop ones as well. Carrots do not like fresh manure: it will make them fork.

Early varieties such as Early Nantes or Primo are sown in spring, and do not require thinning: they are just pulled as needed for salads, or as a delicious cooked vegetable. Main crop varieties such as Chantenay Red Cored or Flak are sown in late spring to autumn for cropping in succession.

Carrot seed is very fine, and main crop sowings will need to be thinned. As carrot fly can be a problem, it is important to carry out this thinning either while it is raining or immediately before or afterwards: the carrot fly is extremely sensitive to the smell of bruised carrot foliage and will be attracted to the area, and rain helps to 'damp down' the smell and makes an attack less likely. Another precaution is to sprinkle salt around the carrot plants to deter the flies (the original wild carrot grows near the sea and is used to saline winds); this must be

Spring cabbages such as Mastergreen are sown in summer, then planted out in their final positions in autumn. They are then ready to harvest as spring greens the following season.

Summer cabbages such as Primo are sown in late winter to early spring, and transplanted when the soil has warmed up in order to have cabbages ready for cutting in summer and autumn.

Winter cabbages, such as the wrinkled-leaved Savoys, are sown in spring for planting out in early summer. Suitable varieties include January King and Savoy King. They are then ready for cutting in late autumn to winter. Once the cabbages have been cut, a vertical slit in the stalk will encourage a second crop of small heads.

Red cabbage is available in summer – a variety such as Primero – and in autumn, such as Ruby Ball. Cultivation is the same as for other cabbages. It is excellent for pickling.

Organically grown Savoy cabbages.

done in dry periods. The best form of protection, however, is to erect a 60cm (2ft) barrier around the carrot bed. The carrot fly does not fly very high, and the barrier deters it. Sytan is a variety that has been developed for more resistance to it. Horticultural fleece can also be used as a protective cover.

Cauliflowers

Firm, fertile soil. Sunny or slightly shaded position. pH 7.0–7.5

The conditions are the same as for other brassicas. Cauliflowers should be given adequate water when they are transplanted, otherwise they take a long time to establish themselves. Some people even wait until it is pouring with rain before transplanting them, but such dedication is rare!

There are summer, autumn and winter varieties. The first type, such as Mayflower, are sown in the greenhouse in late winter and planted out in spring after hardening off. They are ready for cutting in summer. Autumn varieties such as Snow Crown are sown in spring for transplanting in summer, and are then ready for cutting in autumn. Finally, winter cauliflowers such as Snowball are sown in early summer for transplanting in late summer to early autumn. These need protection: either cloches that should be placed over them in late autumn, or a greenhouse. Some people delay the final planting out until greenhouse tomato plants have been removed at the end of the season, so that the cauliflower plants can take over the vacated bed.

Celeriac

Fertile soil. Light shade. pH 6.5–7

If there is a damp, shady area that is not suitable for anything else, it is worth considering celeriac. This is a hardy plant grown for its celery-tasting root. As germination can be rather erratic, we always found it better to sow it in the greenhouse and then transplant it when the soil had warmed up in spring. When planted out, the seedling crown should be left uncovered. It needs plenty of water and a long season to produce nice, rounded roots. If not ready for harvesting in the autumn it can be left in the ground to over-winter and then dug up as required. Monarch and Snow White are reliable cultivars.

Celery

Fertile soil. Sunny or semi-shaded position. pH 6.5–7.0

This used to be a difficult plant to grow well because there was so much work entailed in ensuring efficient earthing up and blanching of the stalks. The self-blanching varieties such as Victoria are much easier to grow, and while they may not be as big or likely to win prizes at shows, they are more convenient for those who merely want a few plants for soups and salads.

Celery needs an adequate water supply. It is best sown in seed trays in spring, and planted out 15cm (6in) apart in early summer. It is ready for cutting from mid-summer to the first frosts. In mild winters we have had plants of the new red-stemmed varieties such as Pink Blush that continued to grow through the winter until they were cut in spring.

Chicory

Fertile soil. No manure. Semi-shade. pH 6.5

Witloof or Belgian chicory is grown for its salad leaves: these must be blanched (forced in darkness) otherwise they are bitter to the taste. Seeds from a variety such as Zoom are sown in spring and thinned out to allow a single taproot to develop. In autumn, the taproots are dug up and the leaves trimmed off before planting in pots or a bed where light can be excluded. The head of pale yellow leaves is then harvested as needed. Large, upturned plant pots or black plastic can be used to exclude light. The area underneath greenhouse staging is a suitable place.

Sugar-loaf chicory, such as Snowflake, is an easier proposition. Its leaves are so tightly packed that they do not require blanching, and they are grown like lettuces. Red chicory (Radicchio) can also be grown unblanched, particularly with newer cultivars such as Red Treviso or Verona Palla Rosa.

Endive is another salad crop that was traditionally blanched to make the leaves less bitter. Again, the availability of newer varieties such as Sally that are self-blanching makes it more popular in the kitchen garden.

Plastic mesh protection for celery. A similar structure protects carrots from carrot fly.

Courgettes, Squashes, Marrows and Pumpkins

Fertile soil with plenty of water. pH 6

All the marrow family, which includes courgettes, vegetable spaghetti, squashes, pumpkins and marrows themselves, need the same conditions. The large seeds can be sown, narrow side up, in individual pots in the greenhouse in spring, and planted out when all danger of frost is past. They need plenty of water. The marrow family as a whole is rambling, and needs plenty of space. They do well on deep beds with an adequate supply of organic matter. We grow them in this way with nasturtiums and borage planted with them; both of these annuals are said to be good companions for the marrow family, and the nasturtium leaves and borage flowers are good in salads. Nasturtium seeds can also be harvested and pickled as an alternative to capers.

Courgettes such as Ambassador should be picked regularly before they get too big; this will increase the number produced and prolong the picking season. Any that are left will grow into marrows. Squash plants such as Sunburst produce small, flattened fruits that are just right for steaming. Vegetable Spaghetti is well named: cooked whole and then opened, the contents look just like spaghetti. Pumpkins such as Big Max or Mammoth Orange are essential at Halloween, and there is always a demand for surplus ones.

Curly Kale

Fertile soil. pH 7.0–7.5

Curly kale is extremely hardy and is useful as a winter green vegetable. Like all brassicas, it is happiest in an alkaline soil, but is less likely to be affected by clubroot than other members of the family. Sown from spring onwards, it can be thinned out as required, with the thinnings being used in the kitchen. The main plants are then harvested in summer to autumn. Later sowings will provide winter leaves. Reliable varieties include Pentland Brig and Fribor.

Fennel

Fertile, well drained soil but plenty of moisture. pH 6.5

Florence or sweet fennel is a tall, feathery-leaved plant that is grown for its aniseed-flavoured bulb, although the leaves can also be used. As it does not like being transplanted, it should be sown direct in late spring to early summer and then thinned out to around 30cm (12in) apart. Bolt-resistant varieties such as Zefa Fino and Cantino are more likely to produce bulbs than older varieties, but plenty of water is required so that there is no setback in growth. When the bulbs begin to form they can be earthed up and mulched. As soon as they are swollen, they can be dug up and are ready for use in the kitchen.

French Beans

Fertile soil in a sunny position. pH 6.5–7.0

Dwarf, French or kidney beans are excellent vegetable crops, although they dislike the cold and will usually rot if sown too early. We sow ours from spring onwards, in succession, to provide a continuous supply through the summer and autumn; a staggered sowing or planting of four plants per 30sq cm (12sq in) is satisfactory. They need adequate supplies of water, and the soil in between the plants should be forked regularly to stop 'panning' or the formation of a hard crust. Alternatively, a mulch can be placed around the plants. They do not need any form of support unless they are climbing varieties such as Cobra or Blue Lake.

Once ready, the beans should be picked regularly and frequently, otherwise they will get too large and stringy. A surplus freezes well. It is worth concentrating on stringless varieties: as their name suggests, they have less of a tendency to become stringy; they include Delinel and Maxi, a variety that also has its beans above the foliage for ease of picking.

Kohlrabi

Fertile soil. pH 7.5

Kohlrabi is grown like other members of the brassica family, but for its swollen stems rather than its leaves. Sown in spring where it is to crop, the plants are thinned to about 23cm (9in) apart, and harvested when the bulbs have developed. Newer cultivars such as Lanro and Kolpak are recommended, rather than older varieties that tend to become woody if bulbs are left to become larger than tennis-ball size.

Leeks

Fertile soil. Open, sunny position. pH 6.5–7.5

The modern cultivars such as Salsa have longer, white stalks and are hardier than older varieties. There are early, mid-season and late varieties, so planting some of each will provide leeks throughout the winter.

They can be sown direct as soon as the soil has warmed up, but we always found it more convenient to sow them in seed trays; they could then be transplanted into the beds as required. Holes about 15cm (6in) deep and 15cm (6in) apart are made in the soil with a dibber, and a leek seedling is put in each hole. Before planting, trim the roots and leaves to about two-thirds of their original size; this helps the leeks to establish themselves more rapidly. The soil is not firmed around the plants, but is left to settle naturally. Once they are watered in, the leeks soon begin to grow.

Lettuce

Fertile soil. Slight shade. pH 6.5

Lettuce is deservedly the most popular salad vegetable and has a place in every garden, particularly as it can be sown in odd corners to fill up spaces. There are several types: Cos, which has thick, upright-growing and well flavoured leaves; Butterheads, with softer, more rounded leaves; Crisphead which, as the name implies, has crisper, tighter leaves; and Non-Hearting, which produces a rosette of leaves but no heart.

Lettuce can be sown in late winter, under cover; these varieties, such as Little Gem and Winter Density, are then ready for picking in spring. The main outdoor varieties, such as Webbs Wonderful and Iceberg, are sown from spring to summer for cropping in late summer to autumn. For over-wintering under cloches or in a cold frame or greenhouse, the varieties Valdor and Kwiek are reliable. They are sown in early winter for cropping in late winter to early spring.

Salad Bowl is ideal for growing as an edging to raised beds. This has finely divided, attractive leaves that can be picked in great handfuls without uprooting the plant, and is justifiably described as a 'cut and come again' variety. The seed is sprinkled thinly on the soil and then left to grow; there is no need to thin out the plants. The result is a solid bank of leaves that do not allow weeds to grow through, thus conserving moisture for the rest of the bed. Picking can continue right through the growing season, without danger of the plants bolting. In many ways it is the best and most economic of lettuce varieties. For really confined places, Tom Thumb will grow in all sorts of nooks and crannies.

Stem lettuce, sometimes referred to as Celtuce, is grown for its stems that can be cooked or eaten raw as a salad vegetable. It requires the same conditions as lettuce.

Onions

Fertile soil. pH 6.5–7.5

There are several types of onion, so there need never be a time when they are not available. The large bulb or keeping onions are planted as 'sets' (individual small onions) and allowed to grow to full size when they are ready for harvesting. They can also be grown from seed, but are more likely to suffer from soil-borne pests. There are white-, brown- and red-skinned varieties such as Albion, Sturon and Mammoth Red, respectively.

Planted in late winter, the small 'set' onions are pushed gently into the soil so that just the tip shows above the surface. If there is a particularly long 'string' at the top, this is trimmed back in case it attracts birds. It may be advisable to net the bed until the onions have established themselves. Spacing on raised beds is around 10cm (4in) apart in a staggered planting.

There are also hardy cultivars such as Unwins First Early and Express Yellow that can be planted as sets in the autumn. These are then ready for harvesting in spring, in the traditional 'onion gap' when the stored onions from the previous season have finished.

Shallots or pickling onions such as Hative de

Lettuce growing through a mulch cover that keeps down weeds and retains moisture.

Onions growing on a raised bed.

Niort can be planted in the same way as keeping onions. The sets are slightly larger, but instead of each one growing into one large onion, it produces a number of smaller ones in a cluster. Both the keeping onions and the shallots are ready for harvesting when the foliage has yellowed and the onions have dried in the sun. The big onions can then be plaited together to form a string for storing.

In the depths of winter the Welsh onion is a reliable supplier of onion foliage. It is best grown as a perennial in the herb garden. The individual small onions can be uprooted, or the foliage used instead of chives. Another perennial is the Egyptian or 'tree' onion that produces a cluster of small onions at the top of the stem.

From spring onwards, there are also chives in the herb garden. This is a perennial that grows easily from seed, providing green leaves for chopping in salads and stews, but it dies down during the winter. Salad onions, such as Crusader and White Lisbon, are sown directly in the soil in spring.

Finally, garlic such as Marshall's Mediterranean is grown in a similar way to onion sets, individual cloves from the heads being planted in the autumn. During the season, each clove multiplies and produces the familiar garlic head with its white, papery skin.

Parsley
Fertile soil in semi-shade. pH 6.5
Sown in spring in semi-shade, parsley provides an abundance of leaves for use as a garnish, as a bouquet garni herb, or in sauces. There are curled-leaved varieties such as Moss Curled, and flat-leaved ones such as Giant of Italy. With greenhouse or cloche protection, parsley can be sown early or planted there later for winter use. Once the flower heads begin to appear in the following year, however, a new sowing is required.

Hamburg parsley is grown for its large, white roots that are used in the same way as parsnips. All the parsley varieties are popular with chickens and rabbits.

Parsnips
Fertile, well worked soil. pH 6.5–7-0
This is a useful winter vegetable that is delicious par-boiled and then roasted in the oven with butter. It is an under-valued vegetable, less frequently grown than it used to be, possibly because it needs a fair amount of room and has a long growing season. It is best sown early in the season, as soon as the soil is workable.

Parsnips require a deep, well cultivated soil. Our heavy soil with many stones was not to their liking, but once we had changed to raised beds, it was a different story. It is important to sow fresh seed, as older seeds soon lose their viability. They should be just covered with soil, and the seedlings thinned out to around 10cm (4in) apart in a raised bed, or 15cm (6in) on the flat.

Parsnips are ready for harvesting when the foliage dies down, but they need not be dug up immediately: a slight frost improves the taste, so they may be left in the ground and dug as needed. Reliable varieties that have some resistance to canker (brown markings) include Avonresister and Gladiator.

Peas
Fertile soil with plenty of water. Sheltered, sunny position. pH 6–5
Fresh peas are one of the most delicious of vegetables, and are available as early and main crop varieties. Early varieties include Hurst Beagle and Feltham First, while reliable main crop varieties include Hurst Green Shaft and Ambassador. There are also mangetout varieties such as Sugar Snap and Oregon Sugar Pod, where the whole pod is eaten.

Peas vary in height, and those with small growing areas may find it better to grow a tall variety such as Alderman for the main crop; this grows to around 1.2m (6ft). Unfortunately it is not resistant to the disease *fusarium* (pea wilt) that tends to affect older varieties.

We grow only the early varieties because main crop peas are available cheaply in our particular area. Early varieties are sown in autumn, under cloches for over-wintering for an early crop. In early spring, a second crop can be sown as soon as the weather conditions are suitable. Again, cloche protection can be provided. Main crop peas can be sown from spring to early summer.

The seeds are sown in rows or clusters around 2.5cm (1in) deep, and should then be provided with twiggy branches or netting as climbing supports. Birds such as pigeons can be a nuisance, and some protection will be needed until the plants have made substantial growth. A scarecrow is also worth its place near early peas.

Peas, like all legumes, can fix atmospheric nitrogen with their root nodules, so they will not need feeding; however, regular watering is essential, and mulching the soil around them helps to conserve water. Once the pods are full, they should be picked regularly so that further cropping is encouraged.

Potatoes
Fertile soil. pH 5–6
There are early, second early, and main crop

Peas, like all legumes, can fix atmospheric nitrogen with their root nodules. This increases soil fertility for the following crop. Photo: Thompson and Morgan

Seed potatoes being set out on chitting trays to produce shoots before planting.

varieties. It is often better to concentrate on early varieties if space is limited; moreover, these are also less likely to suffer from disease such as blight and potato mosaic virus. Besides, in some areas, main crop potatoes can be bought so cheaply from local growers that it is not worth growing them.

Seed potatoes that are certified healthy are widely available, and can be started indoors by 'chitting': the tubers are placed 'rose' end upwards in shallow trays (cardboard egg trays and boxes are ideal) – the 'rose' end of a potato being where the 'eyes' or shoots are situated. Rubbing out all but three of the shoots will provide fewer, but larger potatoes.

The sprouted tubers should be planted in a bed that has not been limed, as soon as the soil has warmed up in spring (lime in excess of their needs will cause scabbing of the skins). Cloches or agricultural fleece can be used to provide early protection.

A wide drill, 15cm (6in) deep, is made with a hoe so that the tubers can be planted in rows 50cm (20in) apart, with 30cm (12in) between each plant. Well rotted manure can be placed in the drills, with the potato tubers on top. We always place freshly cut leaves of comfrey in the trenches: these rot down quickly without nitrogen depletion from the soil, providing a natural potato fertilizer.

Once planted, the tubers are covered with soil and, as the first stalks appear, soil is drawn up around them. This 'earthing up' prevents the 'greening' of potatoes as they are produced. Green potatoes are poisonous. Alternatively, they can be covered with black plastic with holes provided as soon as the shoots emerge from the soil. Potatoes can be lifted once flowers have opened on the plants. Early varieties should be used immediately, but main crop ones can be stored in sacks for winter use: be very careful to exclude light when doing this, otherwise they will turn green.

A recent innovation has been the introduction of potato plantlets grown by tissue culture, as distinct from tubers. These are disease free, and when received should be planted in pots, in protected conditions. Once hardened off they can be planted outside.

Potatoes can also be grown in containers, a procedure that is popular with those who wish to produce new potatoes for special occasions such as a Christmas or Thanksgiving dinner. A couple of chitted tubers are placed on a 10cm (4in) depth of good soil or compost in the container, and are covered with compost. As the stems appear, further compost is added to keep the lower part of the stems (but not the leaves) covered. This procedure continues until the container is filled.

The early variety Epicure has a superb flavour, while the main crop Cara has some resistance to blight. There are many other varieties grown for their culinary and disease-resistant qualities, so it is worth studying the seed catalogues. Old or heritage varieties are also available.

Radish

Sun or slight shade. pH 6.5
These can be grown anywhere where there is a spare patch of ground; it is merely a matter of sowing a pinch of the seed every week, so that not all the plants are ready at the same time. They are quick growing, and must be picked before they bolt and become woody. Even if they have been left too long, they need not go to waste as far as livestock are concerned, for both rabbits and goats will welcome them in their diet. Reliable varieties are French Breakfast, Cherry Belle and White Icicle: the latter is pure white and forms a long, tapering root, making a novel addition to salads. China Rose is a variety for sowing in mid-summer or early autumn for winter salads. It produces thick radishes, up to 5cm (2in), and provides a welcome taste of summer when winter arrives.

Runner Beans

Fertile, moisture-retentive soil. Sheltered, sunny position. pH 6.5–7
This is a vegetable that is most economical of space, and as long as adequate supports are provided for it, it will produce a good crop of beans for day-to-day use, or for the freezer. The best crop is produced by digging a trench the depth and width of a spade spit in winter, and placing compost or manure in it before replacing the topsoil.

The large seeds are either sown direct in the soil, as soon as it has warmed up in the spring; or they can be sown in pots for subsequent transplanting. As climbers, runner beans need sturdy supports: these can be poles with canes arranged in rows between them, with a horizontal crosspole to keep the structure stable; or a wigwam structure of canes can be used. Adequate watering is essential, and mulching is important, for any degree of drought will cause a premature flower drop. When the plants have reached the top of the canes, they can be pinched out at the top to encourage more side shoots.

We always grow sweet peas with our runner beans. It is nice to be able to pick a bunch of fragrant sweet peas at the same time as the beans. They should both be picked frequently to encourage further growth. The annual herb summer savory is also a good companion plant for beans. Modern stringless cultivars such as Polestar and Desirée are preferable to older varieties that have a greater tendency to become stringy. Desirée has white flowers rather than red, and a mixture makes a colourful, temporary hedge with the sweet peas.

Salad Leaves

Fertile soil. Open situation. pH 6.5
Lamb's lettuce or Corn Salad, sown in high summer, will provide winter salads. Other greens that can be

Runner beans can grow high, so they need plenty of support.

sown in vacant spaces at this time include winter purslane (*Claytonia*), Rocket, and American Land Cress. The leaves can then be picked for winter or early spring salads.

Salsify and Scorzonera

Light soil. No fresh manure. pH 6.5
Salsify such as Mammoth and scorzonera such as Habil are grown for their roots, although the former is a biennial while the latter is a hardy perennial. If necessary, scorzonera can be left to grow on for a second season before harvesting, but salsify must be gathered in the autumn after a spring sowing. If a few plants are left in the ground, however, the flower buds that form in the second season can be used in a similar way to asparagus shoots, as long as the stems are picked before the flowers open. If required, salsify roots can be lifted, and after trimming off the leaves, can be planted upright in boxes for blanching. The resulting chards will be ready in early spring if light is excluded and the boxes are placed in a protected position.

Seakale

Sandy but fertile soil. Open, sunny position. pH 7
Another hardy perennial, seakale is a native of Britain that is grown for its blanched stems. It is a deep-rooted plant that can be grown from seed or from root cuttings called 'thongs'. Seed germination can be slow and erratic, so it is better to choose a modern variety such as Angers that is easier to grow.

Thongs should be planted 60cm (2ft) apart, with the first crop being taken two years after planting. In spring, the plant is covered with an upturned bucket or similar container so that light is excluded. The blanched shoots are then cut when they are about 30cm (12in) long.

Spinach

Fertile soil. Slight shade. pH 6.5–7.5
This vegetable is superb when cooked with a few teaspoonfuls of water only, then put in a blender with butter, salt and black pepper; it is a good source of fibre in the diet. It will do well in a slightly shaded area, such as between taller crops. Summer varieties such as Trinidad can be sown, in succession, from early spring onwards, but must have plenty of water to prevent bolting or running to seed.

Spinach or leaf beet such as Perpetual Spinach is easier to grow, and can be sown in succession from spring onwards, with later cropping lasting through to the following year in mild winters. Swiss chard, sometimes called Seakale beet, can also be used as spinach, and later sowings can be harvested in winter.

There are attractive red varieties such as Ruby Chard, as well as the normal cultivars such as White Silver. The stems can be eaten as well as the leaves.

Sweetcorn

Fertile, well drained soil. Warm, sheltered position. pH 6
Sweetcorn is fairly tender. It is best sown in pots and transplanted into beds when the soil is warm. Planting in blocks or in a staggered pattern, with plants 15cm (6in) apart, makes wind pollination easier. The best cultivars are the 'supersweet' type such as Sweet Nugget and Sweet 77. These, as the names imply, are much sweeter than older varieties. There are also mini varieties such as Mini Pop that can be grown in confined areas, or even deep patio pots.

Mulching with compost or black plastic helps to conserve moisture, and also keeps the soil warm. The cobs are ready for harvesting when the tassles at the ends of the ears turn brown. They should be cooked immediately before they lose their sweetness. They are also useful for feeding to livestock.

Turnips and Swedes

Fertile soil. Cool, moist conditions. pH 7.5
Turnips and swedes are both members of the brassica family and so need similar conditions. Swedes are extremely hardy and are normally grown for winter use, while turnips can be sown from spring onwards for cropping in succession. The newer varieties of turnips are fast maturing, enabling cropping to take place as soon as the roots have filled out. Early varieties, such as Snowball, are sown in spring and thinned out as soon as possible for cropping in summer. Some of the Japanese cultivars, such as

Swedes ready for harvesting.

Tokyo Cross, are also fast growing but need to be sown in summer and kept well watered to avoid bolting. Later varieties, such as Green Top Stone, are hardier, and are sown in summer.

Swedes such as the clubroot-resistant Marian are bigger, usually have yellow flesh, and are sown where they are to crop, thinning out to around 23cm (9in) apart; they should be kept well watered. They may be harvested in the autumn and winter, and can be lifted and stored in sacks or in sand.

Herbs

No kitchen garden is complete without at least a small patch of herbs. The culinary herbs enhance a wide range of dishes, while minor complaints such as colds and coughs can often be aided by soothing herbal drinks (although they should not be regarded as a substitute for medical advice). Livestock appreciate small amounts of herbs in their diet. Many herbs are excellent as companion plants, and also for bees; and they can be used in the manufacture of scented articles such as lavender bags and potpourri. Last but not least, they are attractive to look at.

Taller herbs, such as angelica and tarragon, are best planted at the back of a border, with medium growers such as lavender and rosemary in the middle, and the low growers such as marjoram and thyme at the front. Some herbs are invasive – for instance comfrey, mint, balm and horseradish – and are best planted in their own beds where they can be kept under control.

Most herbs can be sown in the spring, while clumps of perennials can be divided in the autumn or winter. Root or stem cuttings with a 'heel' can also be taken in summer and planted in pots, in a sandy compost; if left in a moist, shady position, the cuttings will soon grow roots, and the new plants can be planted out in the autumn.

Herbs that are grown for their leaves, flowers or seeds can be hung in a dry, well ventilated area until they are thoroughly dry and ready for storage. It is also worth experimenting with a microwave oven for drying; and some leaves, such as mint, can be quick frozen. Hang flower or seed heads in a perforated paper bag, which will catch the flowers or seeds.

Flowers

Flowers are essential in any garden, whether they are planted en masse in a herbaceous border, interplanted with vegetables, or grown in containers such as hanging baskets or patio pots. They may be required as cut flowers for sale, or to enhance the home. Whatever the reason, there is a flower for every situation, with the range of plants including perennials, biennials or annuals.

Container-grown kitchen herbs. Photo: Pan

The sage bed in the herb garden. The plants are popular with bees as well as being useful in the kitchen.

Popular Herbs

(A = Annual   B = Biennial   P = Perennial)

Herb	Height	Propagation	Uses
Angelica (B)	Up to 2.5m (8ft)	Sow in summer for following year	Candied stems for confectionery
Balm (P) (Lemon balm)	90cm (3ft)	Spring-sown. Divide roots in winter. Semi-shade. Invasive	Lemon-flavoured leaves steeped in hot water for drinks. Eaten by goats and rabbits
Basil (A)	30cm (1ft)	Sow in pots in spring. Likes sun	Sprinkle on tomatoes. Also for making pesto
Bay (P)	Up to tree size	Rooted cuttings. Sun and protected position	Leaves used in bouquet garni
Bergamot (P)	90cm (3ft)	Divide roots in winter	Infuse leaves for tea
Borage (A)	90cm (3ft)	Seeds itself regularly	Flowers sprinkled on salads or floating in cool summer drinks, but remove sepals from back of flowers
Caraway (B)	60cm (2ft)	Sown in spring. Harvest seeds in second year when they darken	Seeds cooked with cabbage, fruit, in cakes, or sprinkled on bread rolls
Chamomile (P)	30cm (1ft)	Sow in spring	Tea from infused flowers. Eaten by rabbits
Chervil (A)	60cm (2ft)	Sow in spring. Semi-shade	Aniseed-flavoured leaves and seeds for soups, salads, meat and oily fish
Comfrey (P)	90cm (3ft)	Divide roots in winter. Invasive	Leaves for fertilizer. Eaten by goats
Coriander (A)	90cm (3ft)	Sow in spring. Semi-shade	Leaves in salads or Eastern recipes. Seeds in curries and pickles
Dill (A)	90cm (3ft)	Sow in spring	Leaves for pickles, vinegars and sauces
Horseradish (P)	Up to 1.2m (4ft)	Root divisions. Invasive	Roots grated for horseradish sauce
Hyssop (P)	60cm (2ft)	Sow in spring. Cuttings in summer	Can be clipped as hedge. Leaves and flowers in stuffings and with meat and fish. Potpourri plant
Lavender (P)	30–90cm (1–3ft)	Sow in spring. Cuttings in autumn	Clipped as hedge. Potpourri and sachets
Lovage (P)	2m (7ft)	Sow in spring. Root division later	Leaves in salads and soups
Marjoram (P) (Oregano)	30cm (1ft)	Sow in spring. Divide in winter	Bouquet garni herb. Soups and stuffings. Also potpourri
Mint (P) (many varieties)	60cm (2ft)	Root cuttings. Shade. Invasive	Mint sauce, garnish, potpourri. Leaves dry easily
Nasturtium	30cm (1ft)	Sow in spring where to flower. Also climbers	Young leaves in salads. Pickled seeds an alternative to capers
Rosemary (P)	1.2m (4ft)	Rooted cuttings in spring/summer. May need protection in winter	Sprinkled on meat for cooking. Hair conditioner
Pot marigold (A)	30cm (1ft)	Sow in spring. *Calendula officinalis*, not African marigolds.	Petals in salads and stews

Salad burnet (P)	60cm (2ft)	Sow in spring. Divide in winter.	Leaves used in salads. Eaten by rabbits
Sage (P) (also ornamental varieties)	60cm (2ft)	Sow in spring. Cuttings later. Can become straggly unless clipped after flowering.	Stuffings. Gargle for sore throats. Leaves dry readily for winter use
Savory (Summer (A) and winter variety (P))	30cm (1ft)	Sow in spring. Cuttings of winter savory in autumn	Leaves used to flavour meat and sausages Winter variety is a pepper substitute
Sweet cicely (P)	90cm (3ft)	Sow in spring. Divide in winter. Semi-shade	Leaves and seeds with stewed fruits allowing sugar to be reduced
Tansy	90cm (3ft)	Sow in spring. Divide in winter	Used sparingly in cakes and puddings
Tarragon (P)	90cm (3ft)	Plant in spring. French tarragon is better than Russian tarragon, but not as hardy.	Béarnaise sauce, vinegars
Thyme (P) (many varieties)	8–25cm (3–10in)	Sow in spring. Divide in winter. Sunny position. Light soil	Meat, soups and stuffings

- **Perennials** are plants whose seasonal growth continues from the same root year after year. They normally die down in the winter and come back in the spring. They may be hardy, depending on whether they can be left outside to over-winter, or half hardy, in which case they need to have winter protection.
- **Biennials** grow in the first year and flower in the second. Thus if replacement plants are needed, they should be sown every year, otherwise flowers will only be available once every two years.
- **Annuals** are plants that need to be sown every year because they grow, flower and die in the same season. They are described as hardy or half hardy. The former can be sown outside in autumn or early spring, depending upon prevailing weather conditions, but the latter cannot be sown or planted out until the risk of frost has gone.

If space is available, the traditional cottage garden flowers deserve a place. They are attractive to look at, and many are sweet-scented, providing valuable forage for bees and butterflies. Growing flowers for bees provides a supply of food for them over a long period of time, ensuring that there are no 'hungry gaps' in the season. In return, the grower has good pollination for orchard fruits. Butterflies also relish the colourful, scented flowers.

If flowers are to be cut for the home or for sale, it is best to concentrate on bunches of mixed flowers: this means that you can make up bunches from whatever happens to be available at a particular time. The important things to bear in mind are to ensure a steady supply of flowers for as long a season as possible, and to grow foliage plants as well. The list of possible flower plants and varieties is endless, so it is a good idea to visit gardens and flower shows and to study the seed catalogues thoroughly before making a choice. It is also worth checking which varieties have been given a *Fleuroselect* category for excellence, as these will tend to be less straggly than some older varieties. Dwarf varieties are suitable for growing in containers such as patio pots.

Container Plants

One of the most popular ways of growing plants is to utilize containers such as patio pots, troughs or hanging baskets. The advantages are that no ground preparation is necessary, little or no weeding is required, and a wide range of situations and odd corners can be used, which might otherwise be bereft of plants. Perennial container plants can also be taken with you if you move house. There are also disadvantages, such as the need to provide sufficient plant food in the compost, and to ensure that watering is regular, for the containers will dry up much more quickly than a bed.

There is a wide range of patio containers available, in wood, plastic, terracotta clay, concrete, fibreglass, reconstituted stone and glazed earthenware: in fact, anything that will hold compost and allows drainage can be used to grow plants. When put on the patio, they should ideally be placed on

Cottage Garden Bee Plants

Alyssum, *Lobularia maritima*	Cornflower, *Centaurea* sp.	Mignonette, *Reseda odorata*
Arabis, *A. caucasica*	Forget-me-not, *Myosotis* sp.	Poached egg plant, *Limnanthes douglasii*
Aubretia, *A. deltoides*	Honesty, *Lunaria biennis*	Primrose family, *Primula* sp.
Balm, *Melissa officinalis*	Honeysuckle, *Lonicera periclymenum*	Sage, *Salvia officinalis*
Bluebells, *Endymion* sp.	Hollyhock, *Althea rosea*	Self-heal, *Prunella vulgaris*
Borage, *Borago officinalis*	Hyssop, *Hyssopus officinalis*	Sunflower, *Helianthus annus*
Buddleia, *Buddleia* sp.	Ice plant, *Sedum spectabile*	Sweet violet, *Viola odorata*
Candytuft, *Iberis sempervirens*	Lavender, *Lavendula* sp.	Teasel, *Dipsacum fullonum*
Catmint, *Nepeta* sp.	Lobelia, *Lobelia erinus*	Thyme, *Thymus* sp.
Clover, *Trifolium repens*	Lupin, *Lupinus polyphyllus*	Valerian, *Centranthus rubra*
Crocus, *Crocus* sp.	Michaelmas daisies, *Aster. novi-belgii*	Wallflowers, *Cheiranthus cheiri*

Many varieties, such as this Mendlesham Minx pink, are ideal for growing in containers. Photo: Mills Farm Plants

An original way of growing dwarf varieties of vegetable in containers.

blocks or short supports to keep the drainage holes clear of the ground. A layer of crocks or stones is placed in the bottom of the container, and it is then filled with compost before planting up.

Hanging baskets are available as galvanized steel baskets, or as plastic, wooden or cellulose fibre constructions. They need liners to prevent undue water loss: these include sphagnum moss, coir (coconut fibre), bitumen paper, foam plastic, plastic sheeting, cellulose fibre or recycled wool. Those with sheep will find an excellent use for those bits of fleece that are normally discarded.

An easy way of filling and planting a hanging basket is to place it on top of a large pot or bucket. The liner goes in first, then the compost, and water-holding polymer granules and slow-release fertilizer granules can be mixed into the compost at the same time.

A wide variety of either soil-based or soil-less composts is available in garden centres. The former uses sterilized topsoil with peat, sharp sand, fertilizer and lime added, although the last is omitted from composts for lime-hating plants such as heathers. Soil-less composts are based on peat or coir (coconut fibre).

Many people buy bedding plants to plant up their containers. These are widely available either as seedling plants that can be pricked out and grown on in protected conditions until they are ready to plant out, or as ready-to-plant 'plantlets'. Alternatively, they can be grown from seed in a greenhouse. Some years ago, a farmer in our area started a hanging basket business, with customers bringing their baskets back for replanting every year. It has proved to be extremely successful.

A great variety of flowers, herbs, vegetables and even trees can be grown in containers. A visit to the local nursery will soon establish which ones are suitable.

Irrigation

The best times of day to water are early morning or late evening; at any other time during the day there is a risk of evaporation and damage to foliage by scorching. Also, it is better to give a thorough soaking on an infrequent basis, rather than frequent light sprinklings. If using a hose, make sure there is a valve fitted to the outside tap, to prevent back-flow of water into the mains supply (this is required under watering by-laws in Britain).

Too much pressure from a hosepipe can damage plants and disturb soil, so either reduce the flow, or try one of the new aerator flow fittings that mixes air and water, to produce a high volume, but gentle flow. Sprinklers are strongly discouraged by water

It is necessary to fit a valve to an outside tap, to prevent backflow of water from a hosepipe. Photo: C. K. Tools

companies as being the most wasteful amongst garden watering equipment.

Not all fruit and vegetables need regular watering. Generally speaking, the most important time to add extra water is when fruits are swelling. New plants should be watered thoroughly after planting, and kept moist until they have become established. Containers and hanging baskets dry out very quickly and may need watering as much as twice a day. Baskets lined with a thick capillary matting can hold significant amounts of water, and also provide a useful reservoir for the roots. As mentioned earlier, water-storing polymer granules are also available.

Applying a mulch around plants helps to conserve moisture and prevents hard cracking of the soil. If drying winds are a problem, set up temporary windbreaks, using a heavy duty netting, to reduce wind speed. Fine shade netting can also be used in areas such as open patios, where a scorching sun can damage plants.

Rainwater can be collected by using water butts to capture run-off from gutterings on buildings, sheds and greenhouses. Many come complete with a

Guttering and downpipes can be used to provide water in butts.

Using a seep hose is an economical way of irrigating only the areas that need it.

concave lid that directs every drop of water into an inlet. Diverters are also available to fit down-pipes so that the water goes into a butt; if there is too much rain, an automatic overflow facility sends the water down the drainpipe. There are also complete kits to divert bath water or water from a sink into the garden or butt; these have a valve so they can be turned on or off as required.

Alternative and more efficient methods of watering may be via a seep hose, or a trickle irrigation system. This method applies water only to where it is needed, namely to the roots of plants, thereby avoiding wastage. This can be done by using an irrigation system coupled with a timing device or water computer. A well designed system, once set up, will save a great deal of time as well as water.

Seep Hoses

The simplest method of applying water close to the plants that need it, is to lay out a seep hose. This consists of recycled rubber granules, and is made in such a way that it leaks water along its entire length. It is the same size as ordinary garden hose and so can be connected easily, and there are corner and joint connectors to facilitate layout over a wide area. Seep hoses can be laid on the surface or buried beneath the soil to deliver water to the plant roots without losses by evaporation. They can be placed between rows of vegetables, along centres of raised beds, in herbaceous borders, or by newly planted trees and hedges, and can be fed from the mains or from butts equipped with taps.

Micro Tubing

An alternative to a seep hose is to use an ordinary supply hose together with micro tubes punched into it to supply individual plants. Micro tubing is 4 or 4.6mm (⅛in) in diameter, and can be used to feed individual plant roots, or be fitted with emitters on the surface. There are different kinds of emitters, depending on the plants: examples are drippers, adjustable drippers, small micro-jets or mist nozzles; they can be held in the right position with stakes and clips. This is a useful system for watering permanently positioned hanging baskets, plants on a patio, or a greenhouse for individual plants.

Capillary Watering

This is a method of irrigation for the greenhouse. Basically, the plant pots stand on a rot-proof capillary mat that has a high water-holding capacity. The supply of water can be automated with a device containing a ball valve. It can be operated with the mains supply, or from an overhead tank, but it is essential that the staging surface is completely level. Either a header tank can be bought complete with fittings, or any large plastic container can be adapted.

Drip System

A drip system is also appropriate for greenhouse use. It delivers water to individual plants, or to the surface on which they are standing, so could be used in conjunction with capillary matting. A variety of automatic drip systems is available. Water is supplied either from the mains using a pressure reducer, or from an overhead tank.

If you have a large greenhouse containing plants with a range of different watering requirements, do not despair: it is possible to buy a control box that will provide a range of unique options for between six and sixteen watering stations. A solenoid valve is required for each zone.

Misting

Misting is beneficial to most greenhouse plants. The process can be automated by using a remote leaf sensor that is triggered by evaporation of the surface water to turn on the misting or watering equipment. The length of the mist burst is adjustable, and most systems have a manual override option. Mains electricity and a solenoid valve are required.

A water computer can be incorporated into any irrigation system, whether for indoor or outdoor use. It is a relatively low cost item, and is available from garden retailers. The timer is fitted just below an outside tap outlet, and is easily preset via a small keypad and LCD display, to key in appropriate instructions. It runs on batteries, which would normally last about a year. It is completely flexible, and can be adjusted for winter and summer, programmed to come on during the night if required, and will keep the garden watered even during holiday periods.

A pump is also useful for pumping water to where it is needed, for example to transfer water from a rain butt to a header tank. Small solar panels are now available that not only power small water systems such as this, but can also control ventilation in greenhouses.

6 Protected Cultivation

For garden best is south south-west.

(Thomas Tusser, 1557)

No climate is ideal. The growing season is dictated by the amount and duration of light, warmth, rain and other climatic factors. These are not always available in the balance that we would like, but fortunately the growing season can be extended by the use of protected cultivation. Options include the use of a conservatory, greenhouse, polytunnel, cold frame and cloches.

The Conservatory

There is a large selection of conservatories, of all shapes, sizes and prices, and they do undoubtedly add to the value of a house, providing a congenial extra room in which to enjoy the winter sunshine.

After purchase, the biggest cost associated with glass buildings is that of heating them, although lean-to structures can be heated by an extension of the house central heating system. (Reference has already been made, in Part 1, to the use of

A conservatory helps to conserve warmth in the house, as well as benefiting from the house central heating system. Photo: Appeals Blinds

conservatories in providing warmth for the house.) Many conservatories are now supplied double-glazed as standard, and this is undoubtedly the best option; strengthened glass is a better and safer option than normal horticultural glass. Moreover, single-glazed structures may suffer not only from heat loss, but also from roof condensation and drips. The newer types of glass that darken or lighten in response to outside conditions are useful, albeit expensive. However, glass is not essential for all parts of a conservatory: the use of polycarbonate provides a material that not only serves as an immensely strong roof (the window cleaner walks on ours), but also enables curves to be incorporated in the walls, if required.

The use of blinds is another option, and they are a good idea in addition to existing insulation or double-glazing, as well as providing essential shade in summer. Blinds are available in a variety of materials: reeded wood, woven PVC-coated fibreglass, flame-resistant cotton and polyester. Depending on the type of building, they can be fitted as roller blinds, pleated, retracted folds or vertical Venetian. There is also a choice of operating systems, from simple hand-pulled cords, hand-held wands for higher blinds, and motorized systems. The most sophisticated respond automatically to external conditions, so the blinds are drawn when the sun is too glaring, for example. Many house, conservatory, greenhouse and garden systems can also be adapted and controlled from a household computer, although the technology for this is not yet widespread.

Greenhouses

Like conservatories, greenhouses are available in a wide range of sizes, shapes and materials. The framework might be timber, steel or aluminium, while the panels are glass, polycarbonate or plastic. The structure may have vertical walls with a span roof, or it may be a lean-to against a wall, or be domed or polygonal in shape. Whatever type it is, the needs are the same: adequate heating, growing area, ventilation and watering.

Even the smallest garden usually has room for a greenhouse. Photo: B&Q

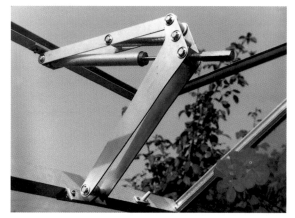

Automatic vent window opener in a greenhouse. Photo: Bayliss

The cost of heating a stand-alone greenhouse in winter, or even one that is against a wall, is generally too high for most people. It may be better to think of it as a 'cool greenhouse', a means to extend the growing season because of its ability to absorb solar radiation. In winter, if it is insulated with 'bubble' plastic it may be possible to maintain the temperature above freezing point. In this way, many patio plants will survive the winter, and bedding or other half-hardy plants can be given an early start.

The most effective way of providing heat is with soil-warming cables, as long as electricity is available in the greenhouse; these are cheap to run and simple to install. The heat is provided underneath the plants, exactly where they need it, and plastic sheeting, or propagator tops for smaller plants, can then contain the warmth in the area around the plants. A properly insulated and earthed cable is essential, as well as a circuit breaker for absolute safety. A good electromechanical thermostat will control the temperature to within one degree.

In summer the temperature can get too high, and shading with roller blinds or by painting on some glasshouse shading may be necessary, as well as ensuring that ventilation is adequate. A thermostatically controlled extractor fan can be used, but a less technical system, which is nevertheless extremely effective, comes in the form of an automatic vent opener. A plug of mineral wax expands with heat, operating a piston, which causes the roof vent to open via a balanced aluminium link. As the wax contracts when the temperature cools, the vent closes by means of a stainless steel spring. Automatic louvre window openers are also available.

Staging, either timber or metal, is essential for holding pots and seedling trays. We utilized part of ours as a soil-heating area. It also had a horticultural bulb suspended above it for lighting.

Polytunnel

Plastic tunnel greenhouses are useful and adaptable structures, capable of being used not only as growing areas, but also as temporary housing for lambing, or as grazing shelters. They are available in various widths and lengths, as required. A typical one would be 5m (18ft) wide, 2.4m (8ft) high and 18m (60ft) long. As the air in a small tunnel heats up more quickly than in a larger one in spring, it is common to have a small, separate structure specifically for seed germination and early plant production. Tunnels can be moved to a new area after a few years, unless a system of plant rotation is used to avoid a build-up of plant diseases. Plastic growing bags are generally used as a convenient way of avoiding this problem.

It is arguably better to buy a small, commercial polytunnel than a larger amateur one because the specifications are usually better. Hot-dipped galvanized steel tubing is the best. The diameter of the hoops should be at least 25mm (1in), while an acceptable gauge for the polythene cover would be 720g. Ideally, the plastic should be of thermal grade to resist the degenerative effects of the sun for several years. Ultra-violet inhibited, anti-fogging film with the addition of ethyl vinyl acetate (UVI, EVA and AF) is a resistant, pliable material that has good light transmission as well as non-dripping qualities. Other types of cover available include a white film for over-wintering shrubs, a dark, insulated one for mushrooms, and heavy duty sheeting for livestock. For the latter, the side sections can be rolled up and replaced with *Nicofencing* secured to wooden lathes: this is a metal mesh that acts as windbreak, ventilation and fence, in our case to confine sheep. After lambing, the polytunnel can revert to horticultural use, with the plastic sides rolled down and the ground turned over. Tomatoes, cucumbers and melons do well after sheep.

A sunny, sheltered and level area of ground is essential for siting a polytunnel. The foundation bars are hammered into the ground, and the hoops inserted. At least three people are needed to put on the cover, so that as two are holding it taut, the third is digging the ends into a trench. A timber batten provides extra stability once the trench is filled. A still, hot day is best for putting on the polythene so that as it cools and contracts, it tightens up. Obviously the door provides ventilation when it is open, but an extra ventilation panel above it is desirable for when it is closed. It is also possible to have side ventilation panels, as well as an anti-aphid screen. Very large structures may need extra roof ventilation. Reinforced PVC mesh can be used for shading. Anti-hotspot, foam-backed tape is available for the areas where the polythene touches the metal so that wear is reduced. There is also special joining tape for repairing tears in the polythene.

Seep hoses are a good way of irrigating tunnel crops where splashing soil onto the foliage is to be avoided. Trickle systems, as referred to earlier, are also appropriate. Some crops require more water than others; also, many people consider that cucumbers and tomatoes should not be grown together because cucumbers require a more humid atmosphere. But with adequate attention paid to watering, ventilation and shading, it is quite possible to grow a healthy range of crops in the same tunnel, while avoiding leaf burn and fungal diseases.

Cloches

Cloches are named after the French word for bell, which the early glass structures resembled. They are essentially miniature greenhouses, and are available in various shapes and sizes, in glass, polythene, rigid clear plastic or polycarbonate. They help to warm up the soil, and provide protection for early or late sown crops so that the growing season is extended. They are available as continuous tunnels for rows of crops, or as individual structures. A continuous cloche should have closed ends, otherwise wind can funnel through, to the obvious detriments of the plants. Those that will fit neatly over deep beds are also available. On a small scale, upturned jam jars

Polytunnels have a wide range of horticultural uses, and can also be adapted as animal shelters. Photo: Clovis Lande

Tomatoes in a polytunnel being grown on straw bales, with early lettuce in front.

Plastic cloches used to protect early crops.

are effective for individual plants. It is worth remembering that while protection of this sort is valuable, the mini-climate underneath may well attract slugs and other pests.

Environmental fleece is also increasingly used, particularly for providing frost protection for early potatoes and other crops. It is made of polypropylene threads spun into a sheet, and with careful use will last for several seasons. The sheet must be weighted down at the edges. Other uses for it include protection against carrot fly, greenhouse shading, winter protection, and protecting fruit bushes from birds. A more sophisticated cover is a crop advancement film (CAF): this is of similar composition but has perforations to allow excess humidity to escape, and moisture from the rain or irrigation system to penetrate.

Whatever type of structure is used, it is important to ensure that adequate ventilation is provided, otherwise the plants may succumb to fungal diseases. Conversely, adequate watering needs to be arranged on a regular basis. With some plastic cloches, one side can be rolled up for ease of access.

Sowing

'Consult the genius of the place' was Alexander Pope's advice to gardeners, implying that the site itself would dictate what could be grown there. The sentiment has much truth, but the question is also easier to resolve with a clutch of good seed catalogues at your elbow.

As a general rule, a seed is planted to a depth of half its size, although fine seed is left uncovered. If planted too deeply, there is a risk of oxygen deprivation and of running out of resources before the shoot can break the surface. A purpose-made seed compost gives the best results, although there may be some who prefer to make up their own. A suitable one would be:

2 parts fine compost (earthworm compost is ideal)
1 part horticultural peat or coir 'peat substitute'
1 part sharp sand or vermiculite

Home-produced compost may need to be steamed to ensure that it is sterile, otherwise weed seedlings may emerge in the germination process. A garden sieve can then be used to ensure the fineness of the mixture.

Seed containers vary depending on the quantity and size of the seeds; for most purposes the seed tray is quite adequate, but various other pots or containers can be used – for instance, large seeds such as broad beans can be sprouted in jam jars before being planted out. Some very fine seed is available in pelleted form, but is more expensive; it is coated with a material to make each seed larger, for ease of handling. Some seeds may be treated with fungicides to protect them against 'damping off', but organic gardeners will wish to avoid these.

Compost is placed in the seed tray and gently firmed. Seeds are then sown according to type, and the instructions on the seed packet: generally fine ones are sprinkled on the surface, while larger ones can be spaced out to avoid having to thin them out later. Fine compost or vermiculite is sprinkled over the seeds to the required depth (unless they are recommended as being left uncovered). Next, stand the tray in water: this allows moisture to be drawn up into the compost without disturbing the seeds. Place a sheet of glass or plastic on top, with a sheet of newspaper to exclude light. Fine seeds are best sown in pots that are then placed in plastic bags until germination is achieved. It is important to check pots and trays every day in case there is too much condensation that might cause 'damping off', where seedlings succumb to fungal disease. Some seeds may require special conditions, such as being subjected to frosty conditions to break the dormancy cycle. This can be achieved by placing the seed packet in the refrigerator for a time specified on the packet. Other large ones with tough skins may need to be immersed in hot water for a short time. Again, the packet instructions are crucial.

Depending on the scale of sowing, a sand bed with under-soil heating cables can be used in the greenhouse; or a plastic propagator with a rigid plastic top where there is access to an electrical point. There are many propagators available, and being able to provide bottom heat does enable a wider range of seeds to be germinated, and to do so outside the normal sowing times.

Once the seedlings have four leaves they are ready to be 'pricked out' (unless they were originally sown in a spaced-out fashion): they are separated and transferred to another container where they will have more room to grow until they are ready for their final planting out in beds or large pots. It is important to keep disturbance to a minimum, and the old saying that 'a plant won't know it's been moved unless you tell it' certainly applies here. Holding each seedling by a leaf and not the stem so as to avoid crushing it, place them in individual holes in the new container of compost, drawing compost gently over the roots. Use a fine spray mist then to consolidate them.

A technique for producing early pea or bean plants for planting out when the weather is suitable, is to sow them in compost, in a section of plastic roof guttering. As soon as they are well grown, the whole unit can then be tipped out and planted as a row section in the vegetable bed. Some gardeners make

Young plants that have been grown from seed being hardened off before being transplanted outside.

use of a 'block-maker', a handy gadget that compresses peat into blocks, with a small dimple at the top for the individual seed. The individual blocks are then placed in a tray, and no thinning out is required. Alternatively, plastic seed trays divided into square compartments can be used.

Sowing many seeds in the greenhouse soon produces a shortage of space, but if the weather is still cold, it may not be possible to transplant them outside. Besides, a process of 'hardening off' is the best way of preparing them for the outside world, using a traditional cold frame, or any of the cloche protection devices mentioned earlier.

When conditions are suitable, the hardened-off plants (or directly sown seeds) can be planted out. Reference was made earlier to the pattern of planting in raised or deep beds, and a handy way of estimating the distance and positioning of plants is to make a wooden frame with some chicken-wire netting tacked on to it; place this on the bed, and then allocate a plant to each space, or to every other, depending on the variety.

Lighting

Using a protected environment for sowing early seeds is one thing; preventing them from growing into pale, straggling plants is another. Low light levels and short days can result in lack of growth or leggy seedlings. To ensure optimum growth, plants need to be provided with about sixteen hours of light a day, with artificial light to extend the natural light (unless of course, sowing is delayed until later in the season). Normal electric light is not suitable; there needs to be a balance of the red and blue bands of the light spectrum: blue is required for strong vegetative growth, while red is required for flowering. Individual bulbs can provide either of these, or a mixture of both. With some systems, it is simply a matter of unscrewing one type of bulb and replacing it with another, as required. On a small scale, one of the best sorts of lighting comes from lamps that combine the qualities of red and blue light. A lamp suspended 90cm (3ft) above the plants would cover a growing area of approximately 90sq cm (3sq ft).

Hydroponics

Hydroponics is a system of growing without soil. The plants are held in an inert medium such as specially manufactured clay pebbles or perlite. Water containing the appropriate nutrients is then made available to the plant. Organic gardeners may regard this as a highly artificial method of growing, but it is an environmentally sound one in that water

A horticultural light bulb in use for early seedlings.

is conserved and there is no leaching of fertilizers into the ground. Hydroponic plants do not need to develop a large root system, and their energies are therefore directed towards upward growth; as a result, their rate of growth is considerably faster than that from normal cultivation. Other positive factors are that no traditional digging, weeding and watering are required, and the danger of soil-borne pests is eliminated. Finally, in a protected environment, and where artificial light is available, plants can be grown all the year round.

There are hydroponic system kits available for the home as well as for commercial growing. The simplest is a passive system, where pots of plants growing in an inert medium are placed on a reservoir tray. Water with nutrients is then added to the tray, and the plants draw it up via the roots. An active system utilizes a pump to make the water available. Delivery is often by means of a nutrient film technique (NFT): plants are grown in plastic troughs that are higher at one end, and lined with sheets of film gathered around the plants. Nutrient solution is introduced at the higher end, and flows down to a collecting tank from where it is then pumped and circulated to all the plants. Another active sytem is the so-called 'flood and drain' method, where the solution is pumped from a central tank to a drainage table where the plant pots are placed, and then allowed to drain away, in a constant cycle.

As far as the nutrients are concerned, they are available as complete solutions or in powder form, and are made up of the complete range of necessary compounds, as well as trace elements such as iron, manganese, copper, zinc, boron and molybdenum.

Plant Propagation

Existing plants can be propagated in order to increase their numbers. There are several methods of doing this: by division or layering, or using runners or cuttings.

Division is the simplest method, where a perennial plant is dug up and then separated with a sharp knife into two or more clumps, each with roots. The best time to do this is in late autumn or early spring.

Some plants, such as loganberries, can be induced to produce roots from the shoot tip, if the shoot is buried in the soil: this is known as layering. Once established, the new root system can be separated and planted elsewhere.

Some plants, such as strawberries, produce runners: offshoots that produce their own set of roots. The strongest of these can be pegged down to the ground and when the root system is well established, the runner can be cut off from the parent plant and planted in a pot or in a new area of ground.

Lastly, propagation can be by taking cuttings: sections that are cut from a parent plant, and planted in sharp sand so that rooting is encouraged. *Softwood* cuttings are the soft and pliable ends of a non-flowering shoot, such as those from fuschias; they should be about 5–8cm (2–3in) long, and the bottom leaves removed from the stem before planting, because it is from here that the roots will grow. *Heel* cuttings are side shoots with a piece of attached bark or stem 'heel'; when the cutting is planted, it is from the 'heel' that new roots emerge. *Hardwood* cuttings are sections of stem that have completed their first year's growth. This type of cutting is often used to propagate trees and shrubs. Sections of stem are cut when the tree or shrub is dormant, and then placed in a large pot or a trench (if outside), and left until they resume growth in the spring.

Protected Crops

The following vegetables are normally grown under protected conditions in the northern hemisphere. In warmer climates they are grown outside, though cultivation requirements are essentially the same.

Tomatoes

This is one of the most popular greenhouse or polytunnel crops. Plants are normally grown as cordons, where a single stem is allowed to grow upwards, the side shoots are removed, and once the appropriate height has been reached, the top is also pinched out. The plants need to be supported, and this is done in a variety of ways: some provide canes to which the plants are tied, but the simplest method is probably to have a horizontal string above the plants, with a vertical string for each plant tied to it, and then anchored to the ground. The plants are then wound around the strings as they grow.

Alternatively they may be planted in a border, though beware if the soil was used for tomatoes the previous year: if so, it needs to be sterilized to get rid of pathogens. A flame gun can be used for this, before and after digging it over.

The most widely used method of planting is to use plastic growing bags. The majority of these are peat based, but it is possible to buy coir- (coconut fibre-) based ones. Once the first fruits begin to form, the plants need feeding. Purpose-made tomato fertilizers are available, but a home-made one, as detailed earlier (page 56), is a perfectly good substitute.

There is an enormous selection of tomato varieties, and it is a personal decision as to which are selected. Some can be grown outside as bushes, where there is no need to pinch out the side shoots – but the harvest is less. In our opinion, many commercial varieties, such as Moneymaker, are tasteless; our personal favourites are Gardener's Delight and Nectar. The former produces small fruits of delicious taste, while the latter is ideal for 'vine-ripened' fruits: they remain firm for a couple of weeks after ripening while still on the truss, and are also well flavoured. Large beefsteak-type tomatoes, such as Big Boy and Dumbello, are popular for slicing.

A lot of problems can be avoided. For instance, tomato leaf mould and botrytis mildews flourish in poorly ventilated greenhouses, while irregular watering can produce split fruit. Too much sun and over-heating often result in 'scald' patches and greenback, where irregular reddening takes place.

Cucumbers

The availability of 'all female' varieties has transformed cucumber growing. With older types it was necessary to remove all the male flowers in order to avoid bitter fruit, whereas nowadays, cucumbers can be grown like tomatoes. Reliable all-female varieties are Futura and Birgit, and for growing in pots (even on a windowsill) the miniature Fembaby is a good choice. Ridge cucumbers, such as the aptly named White Wonder and the round-fruited Crystal Apple, can be grown outside. Gherkins or pickling cucumbers such as Alvin are also grown outside.

Whitefly, red spider mites and aphids can be a problem with cucumbers and other greenhouse crops, although in our experience, inter-planting with French marigolds helps to deter them. There are also biological controls available (*see* Appendix).

Melons

Melons are grown in the same way as tomatoes and cucumbers, but in our experience it is better to restrict the number of side shoots to around four, and train these along horizontal strings. It is also a good idea to hand-pollinate the fruit with a small paintbrush (traditionally, a rabbit's foot was used). They need plenty of water, and the fruit may need some netting as support. Reliable, sweet-tasting varieties are Ogen, Sweetheart and Major.

A commercial crop of tomatoes on the island of Guernsey.

Cucumbers ready for picking in the greenhouse.

Capsicums

Capsicums (peppers) require similar conditions to the crops already mentioned. There are two types: sweet peppers such as Ace and Californian Wonder that produce large fruits, and chilli peppers such as Jalapeno and Apache that have small, tapering fruits.

Aubergines

Aubergines (egg plants) require conditions similar to the above; and avoid erratic watering. It is a good idea to pinch out the top, and to provide support for the side stems. One plant should produce around five good-sized fruits if their number is restricted. Reliable varieties are Black Prince and Slice-Rite.

Mushrooms

Mushrooms have traditionally been cultivated in brick and timber buildings that exclude light. Total darkness is not necessary, however, though a darkened area such as a shed or insulated polytunnel is required. The temperature range needed is 14–18°C (58–67°F), so it is not a practical proposition in winter unless artificial heat can be provided.

It is convenient to construct beds that can be harvested without stooping, and to which there is access from both sides. These are often made of timber, and are bedded up with compost to a depth of 20cm (8in) when it is firmed down. Purpose-made compost is available, but if there is a ready supply of horse manure, this can be used to produce the compost. The manure heap will need to be turned and mixed thoroughly outside, then left for a few days. After this, it is turned again, and the process repeated every few days until there is no longer a

Compost to a depth of 20cm (8in)

Darkened area with a temperature range of 14–18°C (58–67°F)

Growing mushrooms.

smell of ammonia; at this stage it is ready to be transferred to the mushroom house and firmed down. Check the temperature! When it has fallen to 20°C (69°F), it is ready for planting. It should be moist without being too wet.

Mushroom 'spawn' is available from specialist suppliers, and small pieces are inserted in the compost, spaced out 20cm (8in) each way. After about two weeks, white threads or hyphae spread across the surface, forming a fungal mycelium; this should be covered with a 'casing' compost that provides the medium in which the fruiting bodies (the mushrooms) can grow; it also keeps the underlying compost moist. Again, casing composts are available from specialist suppliers, or a suitable mixture would be two parts peat mixed with one part chalk.

The casing is sprayed with enough water to ensure that the bed is damp without being unduly wet. Once the mushrooms appear, they can be picked at the appropriate size, depending on whether small buttons or more open, flat ones are required. Each bed should produce four crops, with an interval of two or three weeks between each one.

Summary

There are many other crops that can be grown in protected conditions, including early sown vegetables for cropping earlier than outdoor varieties, or those that make use of vacated beds in the winter. The greenhouse may also be the place where tender patio fruits such as peaches, nectarines and pot-grown citrus fruits spend the winter.

7 The Orchard

The aspect of an orchard has a bearing on its success. (James Hudson, 1940)

Whilst an orchard of fruit trees needs protection from cold winds, its boundaries should be such that the cold air from frost pockets can disperse – known as 'atmospheric draining'. Hedges have traditionally been used for this purpose because they reduce wind force while allowing air to pass through. In the absence of hedging, netting such as Tensar or Rokolene can be used; this is also permeable, allowing some wind through, but reducing its force and preventing down-spiral. The site should also be well drained and sunny, with a soil pH value of 6.5.

An orchard is an excellent place in which to place beehives, so that good pollination is assured. If it is fenced off, it is possible to graze geese, a few sheep or even a pig under the trees, so that the land has two functions rather than one. The Gloucester Old Spot is referred to as 'the orchard pig' because this is where it was often kept to browse the windfalls. Bear in mind, however, that it may be necessary to place wire-mesh guards around the tree trunks to protect them against damage from livestock.

Top Fruit

The term 'top fruit' applies to fruit trees such as apples, pears, plums or damsons; 'soft fruit' comes from herbaceous plants such as strawberries, or shrubs such as gooseberries. There are several forms in which fruit trees are available to the purchaser:

Standard: A tree with a stem around 1.8m (6ft). It is suitable for a long-term orchard, but not for those anxious to have crops in a short time, or whose growing space is restricted.

Half-standard: This has a shorter stem, usually around 1.2m (4ft).

Bush: This type is smaller still, and is probably the best and most easily managed for smaller gardens. Many varieties can be planted in large tubs.

Cordon: A tree (or soft-fruit shrub) with a single stem that can be trained. For example, cordons can be planted in a row, and at an angle of around 45 degrees to the ground so that a fruiting hedge is made. Shrubs can also be trained upright as double or triple cordons, depending on the number of stems.

Fan-trained: Trees or shrubs that have been trained into a fan shape, normally for growing against a wall or fence.

Espalier: A tree or shrub with one upright stem from which horizontal stems are trained on both sides, at a distance of around 30cm (12in) apart. They are normally grown against a wall or fence. A variation of this is the 'step-over': this has just one stem on either side so that when planted in a row, the plants form a low hedge that can be stepped over.

The height of the tree is determined by the rootstock onto which it is grafted. Grafting is the process of joining a *scion*, or shoot from the tree to be grown, onto a vigorous rootstock that produces better growth. When the tree or shrub is planted, the join is just above the soil level so that suckers or unwanted shoots are not encouraged. Fruit trees are sold ready grafted, but it is also possible to buy grafting kits, together with scions and rootstocks. Some of the rootstocks in common use are MM106 for relatively tall trees, M26 for shorter ones, and M27 for really dwarf trees; the latter can be grown in containers. It is also possible to buy 'family' trees that have several different varieties growing from one rootstock.

A fairly recent development has been the availability of compact column or Ballerina fruit trees: these have a single stem from which very short fruiting spurs grow along the whole length. No pruning is required, and the trees can be grown in tubs, or close together to make a fruiting hedge.

Some trees and shrubs are self-fertile: in other words, their own pollen will fertilize their blossom, as long as there are insects to transfer it from flower to flower. Thus a single Victoria plum or James Grieve apple will produce fruit, whilst other trees

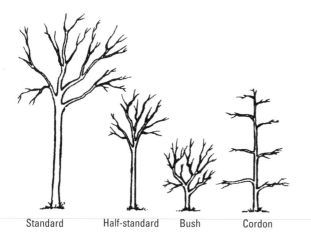

Standard   Half-standard   Bush   Cordon

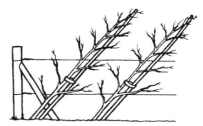

The cordons are tied to canes which are fixed to horizontal wires. Once the cordons are established, the canes can be removed.

Types of fruit tree.

Planting cordons as a hedge.

may require the presence of others in order for fertilization and fruit set to take place. Triploid trees need two pollinators. It is important to check this when buying.

The best time to plant fruit trees and shrubs is in winter when they are dormant, as long as the ground is free of frost. Details on tree planting are given on page 41. Planting distance will vary, depending on the size and type: thus trees on MM106 rootstocks, for example, would be planted about 3m apart (10ft). Whatever the type, it is important to keep a circle around the tree clear of weeds so that its ability to take in nutrients via the outward-spreading roots is not impeded.

Apples

Apples are available as dessert and cooking varieties, as well as early, mid-season or late producers. The ideal is to have a selection of all the groups, while ensuring that they are all matched for pollination purposes. The Bramley is deservedly the most popular of cooking apples, though it has a spreading growth, making it unsuitable for smaller areas. Another disadvantage is that it usually only crops well every other year. A variety such as Arthur Turner has a relatively upright growth and so would be better in smaller gardens – though it is a comparatively early variety, and will not keep for long beyond the autumn. Grenadier is also a good early cooker that has some disease resistance, but again it does not store. Bramley should last until February if stored carefully. For longer storage a variety such as Crispin or Edward VII would be needed.

As far as dessert apples are concerned, one of the

If this apple tree is to fruit, it needs another variety to pollinate it, as well as bees to carry the pollen.

Ballerina apples ready for harvesting. This type requires no pruning.

Pears

Pears flower earlier than apple trees and are therefore at greater risk from frost, although they are more tolerant of heavy, wet soils. A good trio of varieties includes Beth, Conference and Doyenne de Comice because they will pollinate each other. There were no pear trees in our orchard so we planted a half-standard Conference and Doyenne de Comice. In addition we planted six cordons that provided a nice fruiting hedge for the soft fruit area.

Plums

The Victoria plum must be everyone's favourite plum variety: it is a dessert and cooking plum, and bottles well; it is also self-fertile and a regular cropper. Czar is a readily obtainable deep purple plum that is a dual-purpose fruit for cooking and dessert. It is also self-fertile and hardy.

Damsons are popular, not only for bottling, freezing and eating, but also for making damson wine. As we did not have any in our orchard we planted the self-fertile variety Shropshire Damson.

Cherries

Grafted onto dwarf rootstocks, it is possible to grow cherries even in small gardens. Morello and Stella are both self-fertile. Cherry trees do not need pruning, unless it is to remove crossing branches, and this should only be done when the tree is growing during summer to early autumn. It should never be pruned in winter for fear of silver leaf disease setting in.

Soft Fruit

Soft fruit describes the fruit produced by shrubs and herbaceous perennials, rather than trees. They deserve a place on every smallholding, as long as they can be given a well drained site and wind protection. The depredations of birds can also be a problem, and if the scale warrants it, a fruit cage is worth purchasing: this is essentially a box structure made of wood or galvanized metal posts, with wire netting walls; the top is then covered with plastic mesh that can be removed as required (it is essential to allow pollinating insects in). Once the fruit begins to form, the roof netting can be put in position. Alternatively, it is possible to make one's own fruit cage, either with timber supports or interlocking steel poles. On a small scale, a series of garden canes with rubber or plastic ball connectors can be erected around the fruit bushes, and netting draped and pegged down over them.

It was always our practice to put damp

earliest is George Cave. Good mid-season varieties are James Grieve and Tydeman's Early Worcester, while the famous Cox Orange Pippin follows afterwards. Really late varieties for storage are Adam's Pearmain and Newton Wonder.

Cider apple varieties, for those wishing to make their own cider, include Dabinett, Harry Masters Jersey and Yarlington Mill. The best varieties of crab apple are Golden Hornet and Aldenhamensis: these are good for making crab apple jelly, and they are also excellent pollinators for other apples.

The Ballerina apples, that require no pruning (referred to above), are certainly worth considering. There are six varieties currently available: Bolero, an early green eater; Polka, a mid-season red eater; Flamenco, a dark red, late variety, good for storage; Waltz, a late, red-green eater that will store; Charlotte, a Bramley-type cooker that stores well; and Maypole, an early crab apple.

newspapers covered with lawn mowings around the fruit bushes; this acted as a mulch to keep down weeds and retain moisture. After the fruit had been harvested, we let the chickens and ducks have access to the cage, and they did a good job of clearing pests, shredding the newspapers and scratching up weeds.

Most soft fruits require a pH of 6.5–7, with the exception of blueberries which need more acidic conditions.

Strawberries

Luscious, fresh strawberries must be everyone's favourite fruit. It is also an important crop for the pick-your-own market. It is important to obtain clean, healthy and virus-free stock, and to plant in deeply tilled and manured soil, with good drainage.

Young maiden plants obtained from a reputable supplier are planted in late summer or early autumn. It is important to plant properly, ensuring that the roots are adequately spaced out so that they can grow without restriction. Early crops are possible by using polythene tunnel cloches, but it is important to watch out for slugs in this protected environment.

As the plants grow, soil is hoed up around them because they make new roots higher up the crown. This can be done by hand, or mechanically if the strawberries are grown on a field scale.

When the plants flower, straw is placed along the rows so that the fruit will be protected from the soil, and from mud splashed up by the rain. All runners should be removed. The fruit is picked as soon as it is ripe, and is normally sold in cardboard punnets.

Pruning apple and pear trees

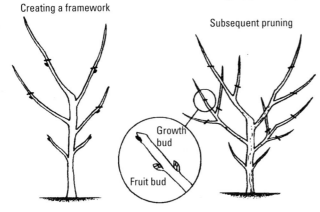

Create a good framework in a young tree by cutting back stems to an outward facing bud. Subsequent pruning is lighter and involves keeping the centre open and maintaining a balance between new growth of stem and fruit production.

Cut out crossing branches to open up centre. Cuts should be slanted in order to shed water. Paint cuts with proprietary sealer to protect tree from infection.

Pruning raspberry, blackberry and loganberry bushes

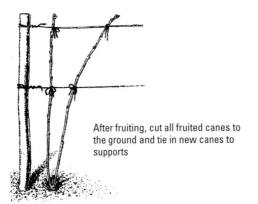

After fruiting, cut all fruited canes to the ground and tie in new canes to supports

Pruning a cordon

Remove crossing or crowded twigs. Cut back long laterals to 3 or 4 fruit buds. Leave shorter laterals

Pruning and training.

Plum trees

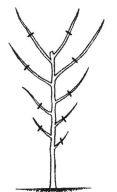

Prune as shown after planting. Little subsequent pruning is needed, other than to remove crossing or overcrowded branches. This should be done in July to avoid silver leaf disease.

Pruning blackcurrants

Open up the bush by cutting out crossing and tangled growth. Remove old fruited wood to make room for new growth.

After pruning

Pruning gooseberries

Remove crossing and tangled growth to keep bush open

Cut out lateral growth on old wood

Pruning and training vines by the double arch guyot method

New growth trained along second wire for cropping next year

First wire

Laterals or side growths cut back after fruiting

Fruited wood with side growths cut back leaving buds for new fruit

Pruning and training. (cont.)

Black plastic has been used as an alternative to the traditional straw for this crop of strawberries on a raised bed.

After fruiting is complete, the straw is often burnt where it stands so that insect pests, weeds and diseases are destroyed, and a small amount of potash is made available for the soil. On a smaller scale, the straw can be removed and the plants trimmed back to about 8cm (3in). A strawberry bed will last for three to four seasons; after that, it is best to make a new one.

Royal Sovereign is an old, fine-flavoured variety, while Cambridge Favourite, a vigorous and reliable grower, follows afterwards. Gorella and Red Gauntlet are mid-season croppers, while Talisman provides late fruits. The best flavoured modern variety, in my view, is Marshmello.

Raspberries

Raspberries are self-fertile and do best in a moist, cool environment. They need well cultivated soil with plenty of organic manure incorporated, and good drainage. Raspberry canes can be planted in winter, and should be cut back hard; the new growth will then be vigorous. The canes are tied onto wires or other supports as they grow, ensuring that they are not damaged and that the fruit can be picked easily. The supporting structure needs to be well anchored and braced in case of wind damage. The simplest construction is made from stakes hammered into the ground at each end of the row, with several wires strung taut between them. Bracing at the ends will provide more stability. In dry periods, overhead spray irrigation may be needed to ensure proper formation of the fruit. The canes are most conveniently planted in rows 1.8m (6ft) apart, with 40cm (16in) between each plant. After fruiting, the fruited canes are cut right back to the ground,

leaving the new canes to provide fruit for the next season.

As with all fruit, it pays to acquire disease-free, certified stock. Varieties that provide raspberries in succession from early summer to autumn are Malahat, Glen Ample and Autumn Bliss. Allgold, as the name implies, is a golden-fruited variety of excellent flavour.

Gooseberries

Gooseberries are often grown on a 'leg': this allows cultivation under the plants, and facilitates the removal of unwanted 'suckers'. They can also be grown as cordons, or trained against a fence to make picking easier. Two- or three-year-old bushes are planted in the autumn at a distance of 1.5m (5ft) apart each way, to allow for subsequent growth and spread. In the first year the leaders, or growing stems, will need to be cut back by half so as to form a strong framework for future growth. Once they are established, the bushes will only require regulatory

With early, mid-season and late varieties, the raspberry season can be a long one.

Leveller variety of gooseberry. By the end of the growing season these fruits will be twice the size.

pruning – this means keeping the centre of the bush open, and removing crowded or rubbing branches.

Careless and Leveller are the two easily available varieties of gooseberry. The former is a culinary type, with fruit which bottles well, while the latter is a dessert variety that can also be used for bottling and freezing. Red varieties such as Martlet are also available.

Blackcurrants

Blackcurrants will tolerate some shade. They are planted as one- or two-year-old bushes, 1.5m (5ft) apart each way. Immediately after planting, the branches should be cut down to one or two buds above ground level to ensure strong new growth in the spring. In subsequent years, pruning should be carried out to remove about one-third of the old, dark-coloured branches so that the bush is kept open and vigorous. Reliable varieties are Baldwin, Ben Lomond and Ben Sarek.

Big bud disease, caused by gall mites, can be a problem because it is also associated with the virus disease reversion, which results in poor crops. Remove any overlarge buds, but if the problem continues, replace the bush. Resistant cultivars such as Ben Connan are available.

Red- and whitecurrant bushes are also available, as well as a newly developed pinkcurrant; these are popular for making jellies and for wine-making. They require the same conditions as blackcurrants, but if space is limited, they can be trained as cordons along wire or on a fence. Varieties include Red Start, White Versailles and Pink Jean. All currants are self-fertile.

Blackberries

Blackberries are a popular bounty of British hedgerows, with people often gleaning the free harvest from roadside hedges. They are also easy to grow in the garden or fruit cage, with the added bonus of having thorn-free varieties such as Helen and Oregon Thornless. A variety that produces extra large fruits is Black Bute.

Blackberries are planted with a similar support system as raspberries. After planting, cut the plant down to around 15cm (6in); as the stems or briars grow, they are trained along the supporting wires. Once fruited, all the fruited briars are removed to ground level, leaving the new ones to provide the next season's fruit.

There are other berry bushes that can be grown, including blueberry, although this needs a damp, acid soil of pH 5. Varieties include Early Blue and Bluecrop. The only pruning required is to remove branches that are crossing in order to keep the bush open.

Other berries include a range of hybrids such as loganberry, tayberry, boysenberry and Japanese wineberry.

Rhubarb

Technically a vegetable rather than a fruit, rhubarb is such an easy plant to grow that it deserves a place in every garden, although it should be remembered that only the stems are eaten; the leaves are toxic. It will grow in quite deep shade. Planted in a perennial bed, the only treatment it needs is the application of compost every winter, after the leaves have died down. Forcing, for producing early stems, is simply a matter of placing a container over the shoots when they first appear through the ground, and excluding light from them so that they grow long stems in a short time. A bucket or wooden box is suitable, and if straw is placed around the container, it provides added insulation against possible frost damage. As referred to earlier, our rhubarb forcer is the cast-iron

Rhubarb forcing with a discarded pot from an old Aga.

pot which came from our old, dismantled Aga cooker. Reliable varieties are Timperley Early and Victoria. When picking, not all the stems should be taken, otherwise the plant is weakened.

Grapes

Vines can be grown in the greenhouse or outside, and there are red, black or white varieties for both environments. It is also possible to have seedless grapes.

For indoor cultivation, a dessert grape can be grown along the roof of a greenhouse, with a support wire set 30cm (12in) away from the glass. The main shoot can then be trained along this, pinching back the side shoots to half size. Once established, and when the vine is dormant, prune back all the side shoots to three buds; this must be done in the dormant period otherwise the cut area will 'weep' sap for a considerable time. When growth resumes and the fruit clusters begin to form, cut back the green shoots to where the fruit cluster is situated. Thinning out about a third of the grapes results in larger fruit clusters. Remove any other green shoots that are not bearing fruit (they will not 'weep'). The shoots are popular with chickens and goats.

For outside cultivation, plants are spaced 1.5m (5ft) apart, in rows 90cm (3ft) apart along a south-facing slope so that they receive the maximum amount of sun. Where machinery is used, the rows may need to be wider apart to allow access. Posts with supporting straining posts are placed at the end of the rows, with two strands of wire running between them, the first at a height of 45cm (1ft 6in), and the second at a height of 30cm (1ft). Planting takes place in winter. The vines are pruned back to three buds immediately after planting, and as growth proceeds, it is tied to the supports. Grapes are produced on the previous year's growth, so no crop can be expected in the first year after planting.

A convenient way of training the vines is to use the 'double arch guyot' method. This involves tying two fruiting canes along the bottom wire, in each direction. The next winter, after harvesting, these canes are cut back and are replaced by two new ones. The fruiting canes are always trained along the bottom wire, while the new growth is trained along the top strand (*see* diagram page 91).

Cluster of grapes ready for thinning out. This entails removing the smaller fruits so that the remaining ones grow larger.

Grape Varieties Suitable for the Northern Hemisphere			
	Red	*Black*	*White*
Indoor	Cabernet Sauvignon (W)	Black Hamburg (D)	Canadice (S)
Outdoor	Triomphe D'Alsace (W)	Baco (W)	Mueller Thurgau (W, D)
	(W= Wine; D= Dessert; S=Seedless)		

8 Pasture and Fodder Crops

Tomorrow to fresh woods and pastures new.

(Milton)

Pasture Management

It has often been said that the most important crop is grass. The best sorts of grasses are those that produce large quantities of leaf growth, or *tillers*, that grow rapidly and recover quickly after cutting or grazing. Apart from the older, traditional grasses in long-established pasture, many new strains have been introduced: some are higher or more erect and are more suitable for cutting for hay or silage; others are close growing or prostrate, and these are generally more suitable for grazing. Strains of particular grasses may also be early or late.

Managing pasture is an ongoing process to ensure a healthy sward, and to maximize the yield for the grazing animals. There are some factors affecting yields that it is not possible to do anything about: for instance, latitude and altitude affect the length of the growing season; sloping fields are less productive than flat ones, because the soil is shallower and there is a greater nutrient loss; rainfall is critical during the summer period when the grass is growing; and finally the type of soil, whether it is clay, loam or sandy, has a direct effect on grass growth under varying conditions. The factors that can be influenced are the soil quality, the sward composition and its management.

Soil

To be available to the plants, the nutrients require a good soil structure and drainage. The humus content improves soil structure and prevents nutrient loss by absorbing the water that holds them. Ploughing up grassland releases the nutrients that benefit subsequent crops, but it also drastically reduces the humus content, which needs to be built up again.

Problems with the soil include poor drainage, indicated by the presence of rushes and sedges, or compaction that produces panning, which is an impervious layer just below the surface. Symptoms

Good pasture is of prime importance to grazing animals such as these Ayrshire cows photographed in Finland.

of this kind require a thorough soil analysis. For grassland, the pH value needs to be between 5.5 and 6.5; a low value can be corrected with an application of lime, which is permitted for organic grassland.

To maximize the yield from grassland, the balance of nutrients needs to be correct. The principal ones are nitrogen, phosphorus (phosphate) and potassium, represented as N, P and K, respectively. If a soil analysis shows any of these to be lacking, an appropriate fertilizer can be applied directly to the field for a quick uptake to restore the balance. An organic farmer cannot use these fertilizers, but legumes to fix nitrogen for the benefit of a following crop can be planted as part of a rotation. Rock phosphate or rock potassium are permitted, but they are slow-acting. Calcified seaweed is a natural product. Poultry manure from a non-battery source and farmyard manure are also useful for restoring the nutrients, but care must be taken to avoid polluting streams and other watercourses under British and European legislation.

Sward Composition

Grassland can be classified into short- and medium-term leys that are ploughed and reseeded every two to five years, depending on the rotation, and permanent pasture. The leys are more productive, but are also more expensive to maintain on a small scale. Permanent pasture can deteriorate over time, unless it has been regularly maintained. Weeds will establish where there are bare patches or over-grazed areas, and thistles, docks, nettles or poisonous plants such as ragwort can appear and seed themselves rapidly if they are unchecked. Ragwort needs to be dug out, but apart from that, most weeds can be weakened and reduced by regular cutting.

Matting caused by a build-up of dead material on the surface is another problem with neglected pasture. A chain harrow used in spring will help to break it up to allow better growth.

It is not necessary to plough up pasture in order to introduce new seed. Scarifying the ground and rolling in the seed in spring can do much to rejuvenate the pasture. The plant species favoured by plant breeders are perennial and Italian ryegrass, which account for 90 per cent of all grass seed sold. Ryegrass has high yields if fed with nitrogen, but the organic farmer will need the right grass to go with white clover: this fixes nitrogen from the air and releases it slowly through the roots. Red clover and lucerne (alfalfa) are also nitrogen-fixing legumes.

Good companions for white clover in the organic sward are as follows. Timothy does well in fertile soils, and is good for late season growth. Meadow fescue does well on poor soil, and although it is closely related to ryegrass, it is less rampant and more flexible, being suitable for permanent pastures and organic systems. Rough and smooth-stalked meadow grass are also suitable for a grass and clover mixture. Cocksfoot does well on light soils that are prone to drought, but it needs to be well grazed otherwise it can increase and dominate the sward. It is a late grass, sometimes being allowed to grow during the autumn to provide useful winter grazing called 'foggage'. For an organic farmer, the ideal pasture is one with a good clover content, combined with less aggressive species.

A selection of other plants may be found in permanent pastures, for example burnet, chicory, yarrow and ribgrass. Their disadvantages are that when baled up in hay they may cause it to heat up, and if eaten in excess by dairy animals, it taints the milk. Their advantages are that they are deep-rooted and therefore resistant to drought, and are a useful source of minerals for grazing stock such as goats, which are predominantly browsers rather than grazers. There are many other common herbs, some of which, such as dandelion, sorrel and plantains, are eaten by livestock. As more farms are going over to organic production, many of the traditional wild flowers of the pasture are also making a welcome return.

Grazing the land hard in the spring stops the grasses from dominating the slower-growing clovers. However, if a hay crop is to be taken, the animals must be removed so that the grass is ready for cutting in early summer. Over-grazing with sheep will deplete the clover because they are particularly keen on it. Knowing how many animals the grass will support is also important, but there are many variables – for instance, productivity of grass depends on length of growth and soil type, as referred to earlier, and it is from these that the potential output of the field can be calculated, taking into account the quality of the sward. However, although these theoretical calculations can be useful, they are more appropriate to large-scale farming. On a small scale, it is probably better to judge by eye and experience, moving the animals when it is obvious that they need new grasses to eat.

Drainage

If pasture is waterlogged it can be improved by drainage, although it may be an expensive option. The best way to establish its condition is to dig a hole about 45cm (18in) deep and see if water collects in it. Pour in an extra two buckets of water! If it has not drained away in half an hour, there is a

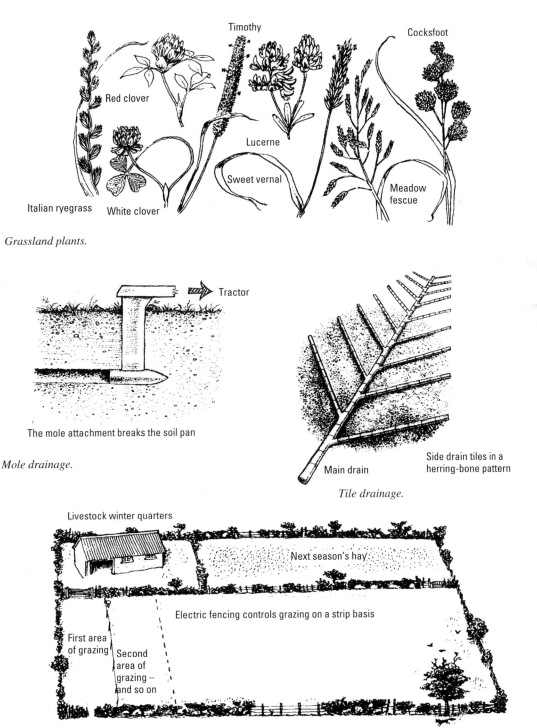

Red clover

Timothy

Cocksfoot

Lucerne

Sweet vernal

Meadow fescue

Italian ryegrass

White clover

Grassland plants.

Tractor

The mole attachment breaks the soil pan

Mole drainage.

Side drain tiles in a herring-bone pattern

Main drain

Tile drainage.

Livestock winter quarters

Next season's hay

Electric fencing controls grazing on a strip basis

First area of grazing

Second area of grazing — and so on

Typical use of pasture.

A flock of sheep effectively confined to one area of pasture by the use of movable electric netting.

problem. If only a small area is affected, such as around field gates where animals congregate, a soak-away can be made, by simply digging a trench and filling it with stones. On a larger scale, a different approach may be necessary – though before making any major changes it is important to check that there are no buried utilities such as electric cables, gas pipes, water pipes, or other pipelines crossing the field. Existing ditches should also be checked, as detailed earlier. If there is an ecosystem of wildlife such as insects, amphibians, mammals and birds taking advantage of wet land, it is a good idea to plan an alternative for them if drainage is to be carried out. This can be done by diverting drained water into a low-lying catchment area; it will soon be colonized and can, if necessary, be fenced off from livestock.

A heavy, clay-based soil may only have become 'panned' or compacted on the surface, in which case a mole plough will break up the pan – a blade attachment is run through the soil about 45cm (18in) below the surface. Applying lime also has the effect of flocculating clay particles so that permeability is improved. If the cause of drainage is more per-manent, it may be necessary to install land drains, though this is the most expensive option; it can be carried out by a contractor if necessary. Land drains are available as flexible piping or as interlocking sections of clay pipe. If the problem is not too serious, one or two pipes running parallel to each

other may suffice; where it is more severe, a herringbone pattern of drains, with side pipes leading away from the main pipe, may be necessary. These will then drain into a ditch or other catchment area.

Controlled Grazing

Between spring and autumn, grass will be available for grazing. The choice of grazing animal will, of course, be dependent upon the individual farmer, but it is important not to let animals have too much lush spring pasture all at once. Scouring can be prevented by putting animals on the grass with hay or straw already in their stomachs.

The best way of controlling grazing is by the use of an electric fence. It is undoubtedly one of the greatest aids of the livestock keeper, allowing him to manage pasture in the most efficient way. As soon as one area of grass is eaten down, the fence can be moved, allowing the livestock to move on. In this way, all the grass is eaten, rather than the more popular grasses being grazed selectively, leaving the coarse ones behind, which is what tends to happen with unrestricted grazing. In winter, animals should usually be left off pasture, unless a 'sacrifice' pasture is used that will be ploughed in the spring.

If the grass year is taken as starting in Michaelmas or 29 September, which is the traditional starting point in Britain, the various activities associated

The Grassland Year

Time of year	Activity	Comments
Michaelmas (autumn onwards)	Plan for next year. Allocate winter quarters and next year's grazing and hay. Lime if necessary.	Do not apply lime and fertilizer at the same time as there will be a reaction.
Winter	Check land drains. Clear ditches. Lay hedges if necessary. Repair fences. Plough land if new ley is to be sown in spring.	Contractor services available.
Early spring	Harrow pasture both ways to pull up dead grass and disperse crusts. Also disperses shallow-rooted weeds and aerates soil. Re-sow bare patches. Harrow winter-ploughed land. Sow new ley and roll it in. Roll permanent pasture to level molehills and consolidate grass. Apply fertilizer.	A contractor can be used. An easy way of sowing is to use a fertilizer spinner and mix the seed with sand to help spread it. May be preferable to sow a new ley in autumn when less bird protection is needed, or it can be sown with a cereal crop which grows more quickly, providing protection for the ley. Rolling usually unnecessary on damp soils, but important on stony ground where stones may interfere with blades during hay cutting.
Late spring	Grazing controlled by electric fencing.	Watch out for signs of scouring from lush grass.
Early summer	Haymaking. Cut hay as it is about to come into flower. Continue to control grazing.	Contractor can be used. Turn and bale as soon as possible.
Midsummer	Take a second cut of hay if ley is good quality.	Alternatively, leave as 'foggage' or late grazing for breeding ewes or other livestock.
Early autumn	Continue controlled grazing if pasture has not been over-grazed and weather is suitable. Plough land for new ley next season. Harrow, sow and roll the new ley field.	New grass sown before end of autumn to forestall later frost damage. Will not usually require bird protection.

with the maintenance of pasture are easier to understand. Michaelmas was the time when harvesting was complete, the winter quarters of livestock were made ready, and there was a pause that allowed for planning and for preparations to be made for the next year. In different parts of the world, the dates will obviously vary, depending upon climatic conditions. The main activities in relation to grassland are detailed in the accompanying panel, but it should be taken as a general guide only, for each farm will vary in its activities, depending upon the weather, the nature of the land, the livestock and the individual farmer.

Hay

Once the grass stops growing, the nutritional value drops correspondingly, and there is no food value to be gained during the winter. For this reason, grass is cut for storage while it still has nutrients. The hay is then available for ruminant livestock, such as cows, sheep and goats, so that they can be fed during the winter.

The choice needs to be made between taking the hay crop or hiring a contractor to do so. He will cut, turn and bale it, and the larger the acreage, the more cost effective this will be. Obviously, for a tiny acreage the cost per contracted bale could work out the same as that of bought-in hay. Some small farmers have their hay cut by larger farmers who take a proportion of the hay bales in return for their service. Another common bartering system is to repay the larger farmer with one's own labour at

Small-scale haymaking in Finland.

harvesting or some other peak activity time.

The best time for cutting is just as the grasses come into flower, but this must coincide with a period of dry, sunny weather. After cutting, the grass will dry on the surface, but it then needs to be turned so that the underside dries. If a tractor with grass cutter and swathe turner is available, it is a fairly easy task. If done manually, a scythe is needed for the cutting and a rake for the turning. This is tough on the hands! A good tip before starting is to place a strip of plaster between the thumb and forefinger, because this is the area that suffers most. Once dry on both sides, the hay is raked into piles from which it can be collected for storage; alternatively, it can be piled loosely onto tripod frames for drying. (In Finland, I was interested to see that they use single upright poles.) Traditional, large haystacks are rarely seen these days, but if they are constructed, they are best covered with a tarpaulin that is roped and staked to the ground.

If done mechanically, and a baler attachment is available, the hay is picked up and neatly packaged into bales, all ready for stacking in a barn. On a small scale, it is possible to produce bales by packing hay into a large cardboard box, in which baler twine is laid. This is then tied when the box is full, allowing the completed bale to be removed for storage. Good quality hay is leafy and green, with a sweet smell.

Silage is grass that is pickled and preserved in anaerobic conditions as a winter feed. Silage-making on a small scale is not recommended. It is difficult to make it without spoilage, and listeria poisoning is a risk associated with inadequately made silage. It should not be undertaken without professional advice.

Poisonous Plants

There are many wild plants that are toxic to animals, although most of the time a well fed animal will not eat poisonous plants. However, every effort should be made to keep them away from grazing areas and from hay meadows where they can become incorporated into hay bales. Ragwort, for example, has been known to poison livestock in this way.

Renting out Land

It is common to find fields rented out to other farmers and livestock owners who need some extra grazing. If you are not using the field yourself, it can

Poisonous plants	
Pasture	Ragwort, bracken, deadly nightshade, woody nightshade, unwilted fodder beet leaves.
Hedgerows	Yew, laurel, ivy berries (the leaves are all right), holly, cypress, green acorns, hemlock.
Garden	Rhubarb leaves, green potatoes, potato haulms, tomato haulms, laburnum, box, foxglove, rhododendron, privet, lily of the valley, brassicas that are in flower, wilting cherry leaves.

produce an extra income from being rented to others. For this it will need to be well fenced, and have water available. If a shelter is also needed and available, this, as well as the water and sound fencing, will make the rental charge higher. If a farmer rents the field on the understanding that the renter is responsible for repairing the fences, then obviously a lower rent will need to be charged.

There are several points to remember if you are considering letting some of your grazing land: the first is that the livestock should be healthy and, where necessary, should have the appropriate 'movement order'. Insist that the animals are wormed before they are introduced onto your land, and ask to see the necessary documents relating to their movement.

Secondly, ensure that the ground is not overstocked, and that you know how long the animals will be there. Thirdly, make it clear that the land is let on the basis that you will accept no responsibility for the livestock. This is important in case of an accident, either to the livestock themselves or to the public in the event of their escape. It is a good idea to have a legal contract drawn up before proceeding.

In tourist areas, there is often a demand for camping space, and this can produce an added income for those who have suitable fields. It is necessary to have permission from the local authority. The local tourist office will also give advice and information, and will put you in touch with camping and caravanning organizations. As long as no more than five motor caravans are parked at any one time it may not be necessary to have permission, but local authorities vary in different areas. It is normal when letting out land for camping, to provide piped water, toilet and litter facilities and level ground. Good access from the road is essential, as well as adequate turning space. It is wise to take out a public liability insurance cover if members of the public use the land.

Fodder Crops

Fodder crops are those that are grown specifically for the feeding of animals. On a field scale, row cultivation is appropriate, with seeds being drilled as thinly as possible. Once the seedlings are up, they can be thinned with a hoe if necessary. Small quantities can be grown in beds, or a surplus from the kitchen garden can be used. Mangolds, swedes and turnips are the traditional fodder crops, although mangolds and swedes have been largely replaced by fodder beet in recent years. Whatever is grown, crop rotation is necessary. Here is one example of many:

Roots	Spring barley or oats, with red clover ley undersown	Winter wheat
Stock graze crop and manure land.	Barley crop harvested. Stock graze on ley.	Harvest wheat, then fallow. Roots after fallow period.

Where animals are 'folded' (allowed to graze the crops in the field), a system of strip-grazing is appropriate. This is where electric fencing is used to control access to one strip of the field at a time. Some fodder crops are also suitable for lifting and storing for subsequent feeding.

Cereals

It is unlikely that anyone would go in for cereal crops unless the necessary acreage, equipment and time were available. Equipment includes a plough or cultivator, with the motive energy to power it. This will be manual labour, tractor or horse, depending on the situation. A seed drill will be necessary (unless the seed is to be broadcast by hand). The crop then needs to be cut with a hand scythe or the hay-cutting attachment of a tractor or cultivator. Threshing is required to separate the grain from the stalks. The grains are then winnowed, where the light chaff is blown away from the heavier grain. All these tasks are normally carried out by a combined harvester whose use, even on a contract basis, would be out of the question for most smallholders.

The most common cereal crops are wheat, barley, oats and maize. These, as well as straw bales, can be purchased as needed from local suppliers. Buying 'off the field' is the cheapest option.

Fodder beet grown for the winter feeding of livestock.

Fodder Crops

Crop	Cultivation	Comments
Carrots	Sown in fine soil. Thin out and lift for storage.	If grown on a field scale, let ruminant livestock and pigs forage on site. Small quantities grown and lifted for rabbits and chickens.
Comfrey	Best grown in a permanent bed.	Cut as hay before leaves become coarse. Several cuts a year possible. Goats and sometimes chickens eat it but it is disliked by many animals.
Cow cabbage	Grow like ordinary cabbages. Can follow early potatoes.	When grazed by cattle, may become mud-spattered. Taller kale may be better. Eaten by ruminants, pigs and poultry.
Field beans (Tic beans)	Winter-sown crop will give heavier yield, with fewer aphid problems.	Ground or kibbled beans provide useful protein for a range of livestock feeds.
Fodder beet	Sow in fine soil in spring. Keep hoed and harvest before frost. Store in clamps.	Wilted tops and roots can be grazed in the field until frosts arrive (unwilted leaves are toxic). Eaten by ruminants and pigs.
Hungarian rye	Sow in late summer as cover crop.	Provides early greens for ruminants and poultry.
Jerusalem artichokes	Plant in late winter when ground is suitable.	Pigs can graze tubers in field. Goats will eat cut-up tubers. For chickens, suspend them whole.
Kale (Thousand headed)	Sow in spring. Keep hoed. Can be strip-grazed in field or fed in stall.	Good source of winter feed for cattle, sheep, pigs, goats and poultry. Beware of over-feeding leading to digestive bloat.
Lucerne (alfalfa)	Good, clean land needed. Can be left for several years or grown as a temporary ley.	Highly nutritious for all livestock and poultry. Can be grazed direct or cut as 'hay' several times a year.
Maize (corn in USA)	Grown in blocks for effective pollination. Needs a mild climate.	Eaten by all large and small livestock.
Mangolds	Sow in spring. Thin out. Store in sand.	Do not feed until mature and touched by frost as they are slightly toxic before then. For ruminants and pigs. Suspend whole for chickens.
Sunflowers	Sow fairly thickly for support.	Nutritious seeds. May need to be chopped for poultry.
Swedes	Sow in spring then thin out.	Suitable for ruminants, pigs. Suspend whole in runs for poultry.
Turnips	Sow in spring or summer.	Suitable for ruminants, pigs and suspended for poultry.

NOTE: No fodder crops should be fed to excess: they need to be balanced with other feeds, such as hay. Root crops for male ruminants should be avoided or fed sparingly because of the danger of urinary calculi.

Part 3
The Livestock

Introduction

In this book it is assumed that all livestock and poultry are kept humanely, on a small scale, non-intensively, and with access to outside free range. They are sentient beings and thrive on good, kindly care, where their natural instincts are respected and catered for. Under existing welfare codes, five freedoms are specified:

- Freedom from hunger by ready access to fresh water and a diet to maintain full health and vigour.
- Freedom from discomfort by providing an appropriate environment including shelter and a comfortable resting area.
- Freedom from pain, injury or disease by prevention or by rapid diagnosis and treatment.
- Freedom to express normal behaviour by providing sufficient space, proper facilities and company of the animals' own kind.
- Freedom from fear and distress by ensuring conditions and treatment to avoid mental suffering.

The keeping of living creatures, on whatever scale, brings responsibilities. They must be looked after every day of the year, including holidays. The reasons why animals are kept should be clear: is it for commercial gain, or is the exercise an interesting hobby? Are they being kept for show purposes, to help conserve endangered old or rare breeds, or to provide for the family?

Keeping proper records makes sense, even if the animals are not kept for commercial reasons. Some, such as the movement on and off site of ruminants, are required whatever the scale. Keeping details of vaccinations and other medical treatments is also a straightforward procedure. Some may complain of bureaucracy, but regulations protect livestock and public alike. Complaints often mask laziness and an unwillingness to move with the times. In the past, there were some who farmed for the subsidies, taking no responsibility for marketing or liaising with consumers, a situation that helped to foster intensive practices and environmental abuses. This book is not for them, but rather for those who are genuinely interested in animals and in the small-scale, local economy. If produce is to be sold, let it be through local outlets where producer and consumer come face to face.

Regulations are referred to in the appropriate chapters, as well as in the Reference section. It is also important to bear in mind that there are zoonoses considerations to be taken into account. This refers to the fact that there are some diseases that are common to humans and livestock, and appropriate care and management are therefore required. Pregnant women or young children, for example, should not help with any veterinary handling activities, such as where animals are giving birth or scouring, because of the dangers of infection.

Finally, nobody should consider keeping livestock without having first acquired practical experience. Going on a course is an excellent preparation for all the various tasks that are associated with animals and poultry. Most agricultural colleges offer a wide range of courses. There are also private organizations and individuals offering specialized tuition. There is no defence for ignorance.

9 Bees

Thus you bees make honey not for yourselves.

(Virgil)

Bees play an essential role in pollinating fruit crops – some bee-keepers derive an income from contracts to supply hives in commercial orchards. Our bees certainly did a good job for us in our small orchard.

Bees are immensely interesting insects, with a well ordered social hierarchy, where each group has its own particular function in relation to the whole hive colony. The queen, the largest insect in the group, is the heart of the colony and lays eggs that provide replacements. The drones, smaller and numerous, are males whose function it is to fertilize the queen during her so-called 'nuptial flight'. The smallest and most numerous bees are the workers, whose job it is to fly to and from the hive, foraging for food in the surrounding area and carrying out all the tasks required in the hive.

Anyone who is allergic to bee stings would obviously be unwise to keep bees. It is a good idea to avoid wearing blue when handling them, and certainly not to inspect the hive when thunder is brewing, for they are sensitive to both. No one should go in for bee-keeping without having attended a practical course on the subject, or having made contact with an experienced bee-keeper who can offer advice. Most local authorities run bee-keeping courses as part of their night school programmes, and agricultural colleges, too, have excellent courses. Many equipment manufacturers also run courses, and will offer a great deal of advice to their customers.

As far as the law is concerned, the hives and their occupants should not be a 'public nuisance' so they need to be sited where people are not likely to be stung. The Roman concept of *ferae naturae* also applies: this means that you own the bees while they are based in the hives, but if they swarm, they are deemed to have reverted to the wild and the swarm becomes the property of whoever takes it.

There are some notifiable diseases that, if suspected, must be reported to the authorities. They include American fowl brood, European fowl brood and varroasis (*see* Coping with Problems, page 12). In the event of infection, no movement of hives is allowed until official inspection deems it to be safe.

The Hive

There are several different kinds of hive to house the bee colony. The National hive is the most commonly used in Britain, but on a worldwide scale, the Langstroth is the commonest. Whichever type of hive is chosen, make sure that the frames are the right size for that particular type. Buying second-hand hives is not recommended, unless it can be guaranteed that they have been cleaned and

Queen whose function is to lay eggs

Worker who carries out all the tasks of foraging and hive maintenance

Drone whose function is to fertilize the queen

Types of bee, showing their relative sizes.

A traditional bee wall with skeps, the forerunners of beehives.

Base to keep hive clear of ground

The hive, and where to site it.

National hive

Roof with crown board underneath

Super with vertically hanging frames of wax foundation for honeycomb production

Queen excluder

Brood chamber with vertically hanging frames for brood comb production

Floor with alighting area and entrance block

sterilized, because the risk of transmitting disease is a real one.

A modern hive is essentially a box with internal ledges on which vertical frames are suspended; it is on these frames that wax combs for egg laying are laid down. For this reason, the box is called a brood chamber, the frames are brood frames, and the wax combs are brood combs.

The floorboard under the brood chamber has one side open to provide an entrance, and this is normally equipped with an entrance block so that the size of the entrance may be altered; for instance in winter it needs to be narrower to protect the interior against bad weather, as well as to exclude larger, robber bees or other predators. In front is an alighting board for returning bees.

Above the brood box is a queen excluder, a zinc or plastic perforated sheet that will allow the worker bees through, but not the queen or the drones; this prevents eggs being laid in the chamber above the brood box. This upper chamber is called a super and is designed to hold the honey frames. There may be one or more, and they are similar in construction to the brood chamber, but normally shallower.

Above the supers is a crown board or inner cover; this has a feed hole, over which a feeder can be placed. Covering the hive is a weatherproof roof, frequently fitted with ventilator grilles at the side. There is also a bee escape, which allows bees to go out but not to get back the same way: the only entrance is at the front via the alighting board.

It is a good idea to have at least two hives so that a spare one is available in the event of swarming, and to cater for future expansion of the colony.

Equipment

Protective clothing is vital for anyone handling bees, with trousers tucked into socks, and gloves pulled over sleeves (a favourite trick of bees is to climb up inside a trouser leg). Bee suits are available that will provide all-over protection. A hat with a veil, giving protection to the face and neck, is the minimum requirement.

A smoker is essential. This has cardboard, oily cotton rags, wood shavings or pieces of sacking ignited inside to produce smoke that can be puffed into the hive. The effect is to provoke an instinctive reaction to feed on the part of the bees so that they are better prepared for an emergency. As far as the bee-keeper is concerned, it makes them much calmer, and so less likely to emerge during hive examination.

A hive tool is a purpose-made tool for levering apart frames that may have become stuck: the propolis produced by the bees tends to make everything stick together, and a good metal hive tool is very efficient at prising frames apart without causing damage.

A feeder is necessary for each hive: this is to provide food for the bees if their stocks have finished and it is still too early in the season for there to be many blossoms about. The feeder is basically a metal or plastic container with a feed access point for the bees.

An extractor will be needed for extracting the honey from the combs, and this is usually the most expensive item. With only one or two hives, it may not be worth buying one, particularly as local bee-keeping societies will frequently hire one out to their members. After straining, the honey needs to stay in a settling tank for a day or two so that any bubbles rise to the surface; a plastic tub with a lid, as is used for making beer or wine, is ideal for this.

Finally, storage jars are necessary to store the honey harvest. Screw-top jars are best, and these, together with labels, can often be purchased through bee-keeping societies.

Siting the Hives

The ideal place to site hives is in a sheltered, sunny position where prevailing winds do not roar around them in winter. They also need to be away from the immediate area where people are likely to be

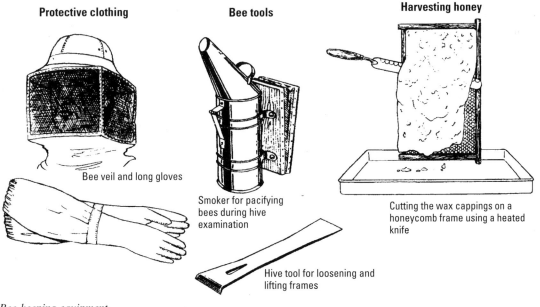

Protective clothing

Bee veil and long gloves

Bee tools

Smoker for pacifying bees during hive examination

Hive tool for loosening and lifting frames

Harvesting honey

Cutting the wax cappings on a honeycomb frame using a heated knife

Bee-keeping equipment.

working or animals grazing; this is because, when bees emerge from a hive they follow a definite flight path, and anyone in the way is in danger of being stung. The way around this is to have a fence or hedge about 1.8–2.4m (6–8ft) away from the front of the hives so that the emerging bees are forced to take a steeply inclined flight path: by the time they are on the other side of the fence, they will be safely above the heads of any unwary gardeners or members of the public. In urban gardens, many people site their hives in a screened-off area at the furthest point from the house. In rural areas hives are placed in orchards if possible, but in moorland areas where there is an abundance of wild heather, hives are often moved there so that the bees can take advantage of the blooms.

We initially sited our hives in the paddock. They were later moved to the orchard after the local hunt went up the lane and the fox veered off across our land. All the hounds followed, streaming over the netting and around the hives and putting the goats and geese to flight. By some miracle, the hives were not overturned, but we decided not to take chances in future and sited them safely in the orchard.

If a hive is to be moved, it is wise to block the entrance with foam rubber and then to tie a rope around the whole structure. Purpose-made 'lock slides' that temporarily hold the various sections together are a useful alternative. A flat-surfaced trolley or barrow is less likely to tip than a garden wheelbarrow, and at least two people are needed to move one hive. For longer journeys, as for example to the moors for bees to work the heather, purpose-made travelling screens made of gauze are available. Journeys should be undertaken at night, or very early in the morning while it is still cool. Once at the new site, the hives should be placed on slabs or some other secure foundation.

Taking Delivery of the Bees

Most large suppliers of bee-keeping equipment also supply bees; there are also specialist breeders who may be contacted through the national bee-keepers' organization. The hive should be got ready before the bees arrive: this entails checking that the brood chamber is placed on top of a floorboard, and that there is a metal queen excluder on top.

Purchased bees will normally arrive in a travelling box, and they will need to be transferred from that into the brood chamber of an empty hive. This nucleus will have a queen with a certain number of worker bees. Once established in the new hive, the queen will lay eggs in the brood combs while the worker bees tend both her and the grubs as they hatch, bringing in pollen, nectar and water to feed them. They fan their wings to create a ventilating draught, and provide guard bees to patrol and protect the hive entrance.

To transfer the nucleus to the hive, first light the smoker; make sure that it is working well, and then put on a bee veil (trying to light a smoker while wearing a veil can be hazardous). Remove the screws that secure the lid of the travelling box, puff a little smoke in, wait a few minutes, then gently lift off the lid. Now lift out the outside comb by holding the projecting ends until it is clear of the box. Carefully turn it to examine both sides. There should be brood combs with eggs and grubs on the frames; these will form the nucleus of the new colony. Place the frame in the brood chamber of the hive, in the same position that it occupied in the travelling box. Repeat the procedure until all the frames have been transferred. If there are any bees left in the box, turn it upside down and shake them into the hive brood chamber; a soft brush or a feather can be used to dislodge any that get stuck.

Any gaps in the brood box should be filled with new brood combs with wax foundation on them so that, as the bees multiply, there is room for them to spread. Put the cover on top of the queen excluder and then replace the roof. Some people prefer to put a super, or honey chamber, on top of the queen excluder at this stage: this is a shallower chamber than the brood box, and holds the combs on which the worker bees will manufacture and store honey. The queen is excluded from this to prevent her laying eggs there. Some people prefer to put the super in position later, but late spring should be regarded as the latest possible time. It will be necessary to feed the bees until they have built up their own stocks.

Water is also important to bees, and it is a good idea to ensure that they have a safe source, in the sun but out of the wind, about 18m (20yd) from the hive. An upturned bin-lid set into the ground and filled with clean water is ideal. A few pieces of bark floating in the water will provide secure landing places.

Feeding

Flowers provide bees with the ingredients for manufacturing food for the colony. Nectar and pollen are carried to the hive by the workers, and then processed by them to produce different types of food. For instance, royal jelly is a whitish, creamy food for the queen bee larva only. The male drone

Planting flowering plants such as this marjoram provides food for bees and ensures that fruit trees are pollinated.

larvae that develop from unfertilized eggs are given less rich food, but enough to provide energy for their nuptial flight to mate with the queen, after which they die. The honey that is produced is a winter food store for the colony as a whole. To save on food, any remaining drones are starved or driven out of the hive in the autumn.

In Britain, the main food sources are white clover and lime trees, although recent years and changing farming practices have seen the decline of clover in favour of rape. Other important plants are heather, fruit blossoms, hawthorn, sycamore, blackberry and dandelions. It is a good idea to plant a range of bulbs, plants and shrubs in the garden, so there is a source of food from early spring to late autumn (*see* list on page 74).

Feeding by the bee-keeper is necessary to make up for any shortage of food in the hive. This may be because the colony is new and has not yet built up its own stock, or because the stock (honey) has been removed, leaving the bees no winter supplies.

To make a sugar solution, dissolve 900g (2lb) of white granulated sugar in 0.5l (1pt) of hot water; allow it to cool, then place it in the feeder. Alternatively, a cake of sugar candy can be used. Finally, it is possible to use a 1kg (2.2lb) pack of white sugar with a hole made on one side into which a small cup of water is poured to stop the sugar from pouring out. Place this, the candy or the feeder on top of the feeding hole in the crown board or inner cover. Feeding should continue until it is obvious that the bees are no longer interested because they are getting enough food elsewhere.

Seasonal Management

The tasks associated with beekeeping vary depending upon the season. All the techniques involved would be demonstrated during a course of practical tuition.

Spring

It is a mistake to open and examine existing hives too early in the season, because this will expose the over-wintering bees to a sudden drop in temperature. Early spring is normally soon enough, and a mild, warm day should be chosen; in these conditions the bees are more likely to be placid, particularly if they have had the advantage of early blossom.

There are two things to establish at this time of year. Firstly, is there a queen, and is she laying? And secondly, is there enough food for the colony? The queen is larger than the workers and has long, amber-coloured legs: if you cannot find her and there are no eggs or brood, it is likely that she has died. If there is a high proportion of high-domed, drone brood combs, it is an indication that she is low in sperms. In both cases, a new queen is necessary.

Early spring can be a difficult time as far as food stocks are concerned. The colony may have come through the winter, but there may not be much blossom about, so if there is any doubt about supplies then the bees should be fed.

If all the combs are full of bees and they are obviously active, there is no need to remove any of the frames unless you find it difficult to see from above whether brood cells have been made. Often it is possible to see this by puffing in a little smoke and then looking down when the bees have cleared to one side. If the brood chamber is full, it is best to provide a second brood box or a shallower super to act as a brood chamber. As bees tend to work downwards in the spring, many people prefer to lift up the original brood chamber and put the second brood box underneath. The drawback is that one has to disturb the bees in order to do this.

Ensure that the new brood box has frames with foundation wax. Ideally, you should use combs that have already been built, because this lessens the workload of the bees at a time of year when they have a lot to do, but plain foundation can be used if made combs are not available. Once the second brood chamber has been placed in position, a queen excluder should be put on top, and a super with frames for the laying down of honeycomb.

If, on the initial spring examination, the brood chamber has plenty of room for expansion, there is no need for a second brood box, and a queen excluder and super can be added straightaway.

Summer

Depending upon the weather and the availability of food, the bees may be so active during the summer season that several supers will be needed. Regular examination will reveal when a new one is needed, and about once a week or every ten days is advisable during the late spring and early summer period. These examinations will also help to prevent swarming.

If new queen cells are formed, the likelihood of the colony swarming is greatly increased. Queen cells can be spotted fairly easily because they are larger than normal brood cells and stick out at an angle to the rest of the comb. Many beekeepers destroy these with the hive tool when they come across them. Specialist bee breeders are trying increasingly to breed strains of bee that have less of a tendency to swarm, and this should be borne in mind when purchasing new stock.

If a new colony is needed, it can be housed in an empty hive consisting of a brood chamber with some old brood comb. Find a frame with queen cells in the existing hive and place it in the new hive; if there is more than one queen cell, just one should be selected, and the others destroyed. There should be quite a lot of bees on the frame, but if there are not enough, more can be shaken off other frames, and foundation brood frames should be placed in the brood chamber for them to work on.

Autumn

After the honey and wax harvests (*see* later) it is important to check the hives and make preparations for the winter. There may have been several consecutive honey harvests, depending upon how good the season was. Wet combs from which honey has been extracted can be returned to the hives for a few days, to be cleaned and repaired by the bees. Then they should be removed, wrapped carefully to preserve them from attack by the wax moth, and stored until needed next season.

Now is also a good time to check the brood chamber, and if possible to find the queen and check that she is healthy. She may have stopped laying at this time, and if there is no indication of brood, it may be necessary to replace her. In the autumn, replacement queens are comparatively cheap and may be

purchased from bee breeders. If the existing queen is healthy but not laying, she can often be induced to start by replacing the central brood frame with a foundation frame. Some of the brood frames may also have honeycombs in them, and this is something to be taken into account when autumn feeding is carried out. Check for the presence of varroa mites, and if they are present, insert Bayvoral strips between the two middle frames of the brood box (*see* Coping with Problems).

Winter

In winter, check that the hive itself is sturdy and free of roof leaks. Reduce the size of the entrance hole in order to keep out larger predators and, if necessary, insert a mouse guard. If high winds are likely to occur, place bricks on the roof in order to weigh it down. When snow falls, it is good practice to place a board in such a way as to cast shade on the entrance: this is to stop reflected light getting into the hive and possibly luring bees outside, where the cold would kill them. In the cold winter months the hives should not be opened because the temperature fluctuations could have disastrous results.

Taking the Honey

It is not a good idea to take the early honey because this will be depriving the colony when it is still expanding and building itself up in the early part of the summer; it is better to wait for the main honey flow that, in Britain, is said to begin with the clover and end with the lime blossoms. However, there is

Drone pupae lifted out on an uncapping fork for varroa examination. Mites are visible as dark brown spots. Photo: DEFRA

obviously a considerable seasonal variation, and local conditions will need to be taken into consideration. In a good year there will be so much honey that harvesting may take place several times.

All the equipment should be got ready first: in addition to an extractor, strainer and settling tank, you will need a sturdy working table, capping knife, trays, and jars with screw-top lids. Also, it should be possible to seal off the room where the work is to be carried out so as to prevent the incursion from outside of any stray bees or wasps.

Honey is ready for extraction when the cells in the supers have been capped by the bees; until this happens, the honey will not be in a ripe condition and will ferment in the jars. The exception is where bees have been foraging in oilseed rape fields, when the honey granulates and sets so quickly that it may even solidify in the frames. A solution is to remove the supers of rape honey when the combs are partially capped, then extract, filter and bottle the honey immediately.

A few days before extracting, lift off the super and place an escape board between it and the brood chamber. This will allow bees to go down into the brood area, but not back into the super, so that in a few days there will be only a few bees left to contend with when the honeycombs are removed. When the combs are taken out they should be replaced with new wax foundation combs so that the bees are encouraged to make more honey. The escape board should then be removed from the top of the brood chamber so that the bees have access to the supers again.

The easiest way to carry the frames is to put them in an empty super, and it is much easier if two people carry it because the weight is considerable. A flat trolley is ideal.

Once inside, in a room where angry bees and hungry wasps cannot follow, the frames are uncapped. This is achieved by standing the frame on a tray and cutting upwards with a capping knife; the frame is then turned, and the operation repeated on the other side. It is then placed in the extractor. Bee-keeping equipment suppliers sell electric capping knives that simplify the operation, but a long sharp knife dipped in hot water will perform the operation satisfactorily.

For efficient separation, a separator is recommended. This works by centrifugal force, whirling the honeycombs around so fast that the honey is driven out and collected in a tank underneath. This has a tap so that the honey can be run off into a settling tank; here, air bubbles have a chance to rise to the surface. Before settling, however, the honey must be strained to remove any wax. Suppliers of bee-keeping equipment stock purpose-made strainers, but it is just as effective to place a large-sized kitchen strainer or straining cloth of muslin or nylon over the top of the settling tank and run the honey in through this.

Leave it to settle for a day or two, ensuring that the tub is covered to keep out dust and insects, then bottle in screw-top jars. If the honey is to be sold, the quantity must be stated on the label, and it must be accurately measured so as not to contravene the weights and measures regulations. It is also necessary to have the name of the supplier on the label. Most bee-keeping societies sell attractive labels to which the bee-keeper can add his own name and the type of honey, for example 'Heather Honey' or 'Local Essex Honey'.

A full National hive super will contain approximately 13.5kg (30lb), and if there are several supers in a good year it will be seen that the yield per hive can be high. Note, however, that the honey that is taken must be replaced by sugar solution, or the colony will die in winter.

Heather honey is far more gelatinous than normal honey, and must be pressed out rather than extracted centrifugally. The combs must be cut out, wrapped in a filtering cloth and then pressed. Heather honey presses are available from dealers.

If honeycomb is to be sold – and this is often double the price of liquid honey – squares of comb are cut out. Special comb cutters are available, or a sharp knife dipped in hot water can be used.

Wax

After the honey harvest, part of the clearing-up operation involved in this sticky process is collecting together all the bits of wax and recycling them. The cappings will still have some honey mixed with them, and this can be reclaimed during the filtering process. To reclaim the wax, put the cappings and bits and pieces in a large saucepan with a little water and heat gently until the wax melts. Put the pan aside to cool: the wax will form a cake and shrink away from the sides, making it easy to remove. It should then be gently washed and stored.

Making foundation for new frames is an obvious use of recycled wax, for the cost of foundation frames is now high. Some bee-keepers send their wax to appliance manufacturers to be made into foundation sheets, and this is more costly than it is to make your own; however, the finished product is of a high standard. Commercial moulds are available if you wish to make your own.

The easiest way to make candles from beeswax is to buy plastic moulds and wicks from a crafts supplier. Place the mould upside down with a wick going through the hole at the bottom, and plug the hole with a small piece of putty. Tie the other end of the wick to a pencil resting across the open end of the mould. Place the beeswax in a water bath to avoid over-heating, and heat until completely melted. Pour the wax into the moulds and stir slightly to ensure that there are no air bubbles, which could cause 'pitting' in the finished product. Ensure that the wick is in a central position, then leave to cool. When quite cold the candles can be removed from the moulds.

Beeswax polish is easily made by melting beeswax in a water bath. Extinguish all naked flames and quickly stir in pure turpentine: ½l (2pt) will be needed for every 450g (1lb) of beeswax. When thoroughly mixed, pour into suitable containers and leave to cool.

Swarming

Bees will swarm for a number of reasons: because the hive is overpopulated; because the weather or forage conditions are not ideal; or sometimes because a particular strain of bee has a genetic tendency towards swarming. Whatever the reason, a new hive should be made ready as before, and the swarm taken from its settling place. The easiest place from which to take a swarm is from a low and accessible bough on a tree, but swarms have been known to cluster on the most difficult places, such as on rooftops.

Collecting a swarm really needs two people. If the swarm is hanging from a tree, for example, it involves one person sawing through the branch on which the bees have landed, or knocking it off, while the second person stands poised below with a box and a white sheet to cover it, ready to take the swarm when it comes down. And watch out, because the weight can be considerable! A fine spray of water directed at the swarm will help to contain and calm the bees.

Once the swarm is taken, the box can be placed on its side on a white sheet, with the other end leading to the alighting board at the entrance to the brood chamber. The bees will then crawl up the sheet in their own time and take residence in the empty hive. A quicker alternative for those more accustomed to handling bees, is to remove the two central frames from the hive brood box and empty the swarm into the space. Replace the frames gently after the bees have moved down so as not to crush any bees. One old custom to encourage swarm bees to occupy a hive, is to rub the inside of the brood chamber with some broad bean flowers. Foundation combs with wax should be made available to them so that they can then start to produce new brood. It will also be necessary to feed them, for they will not have their own stores. Place a feeder with sugar solution over the feeding hole in the inner cover, and replace the roof.

Re-Queening

The queen gives of her best in the first two years, and after that her performance decreases. Commercially, queens are usually replaced every two years. 'Supersedure' is replacing the queen without having the process of swarming, and this can be done either by rearing one's own replacement, or buying in a new queen. If a new queen is to be introduced, the old one must be removed first or the worker bees may not accept her. Introducing a new queen often has the effect of making bees more docile if they have been showing signs of aggressive behaviour.

The queen is marked with a dab of colour, not only to make for ease of identification, but also to indicate her age. There are internationally agreed colours as follows:

Year ending in 0 or 5	blue
1 or 6	white
2 or 7	yellow
3 or 8	red
4 or 9	green

To mark a queen, she needs to be captured in a restraining cage and a small dab of colour applied to her back through the cage's bars. Small pots of enamel paint of the type sold for model-making are suitable.

If a new queen is bought, she is normally supplied in a cage with a plug of candy at the base. The easiest way of introducing her to the hive is to remove the candy plug and push the cage (plug end down) between the two middle frames. After a few days, the cage can be removed.

Coping with Problems

The first essential to avoiding problems is to start with clean hives, equipment and stock. Buy only from a reputable dealer or bee-keeper, and after that it is a case of regular inspections and taking precautions to prevent the onset of diseases. If in any doubt, ask for the Bee Officer of the area or an experienced bee-keeper from the local society to come and inspect the hives and colonies.

Braula coeca

A reddish-brown, flea-like, six-legged insect found on bees; it is comparatively harmless. It is a good idea to be able to recognize it, if only to distinguish it from the much more serious mite, varroa, also reddish-brown but crab-like and with eight legs.

Acarine

A parasitic mite that lives in the breathing tubes of the bee. A light infestation may show no outer signs, but a heavy one may result in bees being unable to fly. If there are many around the entrance of the hive, crawling but obviously unable to take off, put some in a matchbox and send them to the Bee Officer of the area for analysis.

Varroa

Varroa jacobsoni is an external parasitic mite of the spider family. Around 1.5mm in diameter, it sucks blood from bee larvae, as well as clinging tightly to the bodies of adult bees. Detection is a priority. A sheet of paper should be placed at the bottom of the hive, ideally with 3mm gauze mesh just above it. The presence of dead mites amongst the hive debris on the paper indicates the presence of mites higher up in the hive.

Varroa mites are more commonly found in drone brood than in other brood combs. Uncap some of the drone brood cells, and check for brown spots on the white larvae. Removing drone combs as they are capped will help to control infestation. *Bayvoral* strips suspended between the brood combs are an effective control. Ideally, this should be done at the same time as all the other local bee-keepers so there is less risk of spreading the mites from untreated hives to treated ones.

Foul Brood

There are two forms of this disease. With American foul brood the larvae die in the cells after the latter have been capped. The cappings appear shrunken and may be perforated. If a match is inserted into a cell and then pulled out, it will be coated in a sticky ropiness: this is a clear indication of the disease. Alert the authorities who will then destroy the colony and impose a 'movement order'. European foul brood can be recognized by the fact that the larvae die before the cells are capped over; they appear twisted and yellow or brown. Again, contact the authorities, who will order the destruction of the colony and impose movement restrictions.

Chalk Brood

Caused by fungal spores in the food; the larvae become chalk white and mummified. If only a few cells are affected, the condition is not serious. However, if it spreads or continues in the following season, get advice from the Bee Officer.

Sacbrood

A viral disease that kills the larvae after capping has taken place. It is identified by the dead larvae that dry out and resemble scales. It is not usually serious and will often disappear of its own accord. If it persists, it may be worth replacing the queen.

Nosoma

A disease caused by a micro-organism in the digestive tract. It is indicated by brown droppings or dysentery that foul the frames and hive front. If suspected, some bees can be sent for analysis.

10 Chickens

The fowl has followed the treks of man in his wanderings, been arbitrarily transported from one place to another, in which respect its wonderful adaptability to varied conditions has been a factor of the greatest importance.   (Edward Brown, 1930)

Chickens are small, popular and productive, and have relatively modest requirements. They are adaptable and will fit in well in most environments, as long as they are provided with shelter, food and protection from predators. Descended from the Red Jungle Fowl of Asia, *Gallus gallus gallus*, they are perching, pecking and scratching creatures and will not respect flower borders and vegetable beds. Like all birds, they also take frequent dust baths where fine soil is trickled through the feathers, an instinctive pattern of behaviour to help get rid of external parasites.

There are three areas of chicken keeping where it is possible to earn an income from a small flock: the production of free-range and organic eggs; the production of free-range or organic table birds; and the sale of ornamental or show breeds. However, there are many people who keep chickens for the simple reason that they like to do so.

In Britain, there are no restrictions on the keeping of poultry on a small scale, as long as small, movable houses are used. Large, static houses will need planning permission. It is a good idea, however, to check the title deeds, lease, or rental contract of the property in case there are any private clauses that prohibit the keeping of poultry. In places close to urban areas there may also be restrictive by-laws, while the USA and Australia have zoning restrictions.

Site

The ideal area for chickens is one that duplicates the conditions to which their ancestors were adapted. It is no coincidence that many free-range table birds in France are reared in forested areas that provide the shelter, shade and security of trees, with ample scratching areas on well drained ground covered with vegetation.

An orchard, a field with hedges and trees, or even a sheltered garden plot can provide a good environment. Trees, hedges, wall and fences all provide shade, protection from winds, and a sense of security from perceived threats overhead, such as hawks (or aircraft). In an open field, the birds tend to cluster around the house rather than making use of the whole site. The area immediately outside the house then becomes over-used and muddy, leading to a greater risk of disease.

A priority is to ensure that the area is adequately fenced against predators such as the fox. A fence that is 1.8m (6ft) high will deter most foxes, as long as there are no spaces at ground level through which they can crawl. In the case of a particularly determined fox, a further 30cm (12in) of fencing along the top, and angled outwards, will normally be sufficient. The best protection, however, is to use electric fencing, either as permanent protection along the perimeter boundary, or as movable fencing in the immediate area to which the birds are allowed access. In the latter case, electrified netting that can be rolled up and repositioned is the best option.

Housing

There is an enormous range of housing available for poultry, from small houses for a breeding trio of chickens, to houses large enough for a commercial flock. They may be movable or static, have an attached run, or be free standing within a fenced-off area. Some small houses offer attributes such as a solar panel power supply, and automatic watering and feeding that have in the past been confined to large, static buildings. Small flocks for commercial use would normally number anything from twenty-five to 100 birds per house, although larger flocks are seen. It is a good idea to obtain the catalogues of manufacturers, and to go and see as many examples of housing as possible. Plans for building your own poultry house are also available. As a general guide, the flock density inside the house should not exceed seven birds per 1sq m (10sq ft).

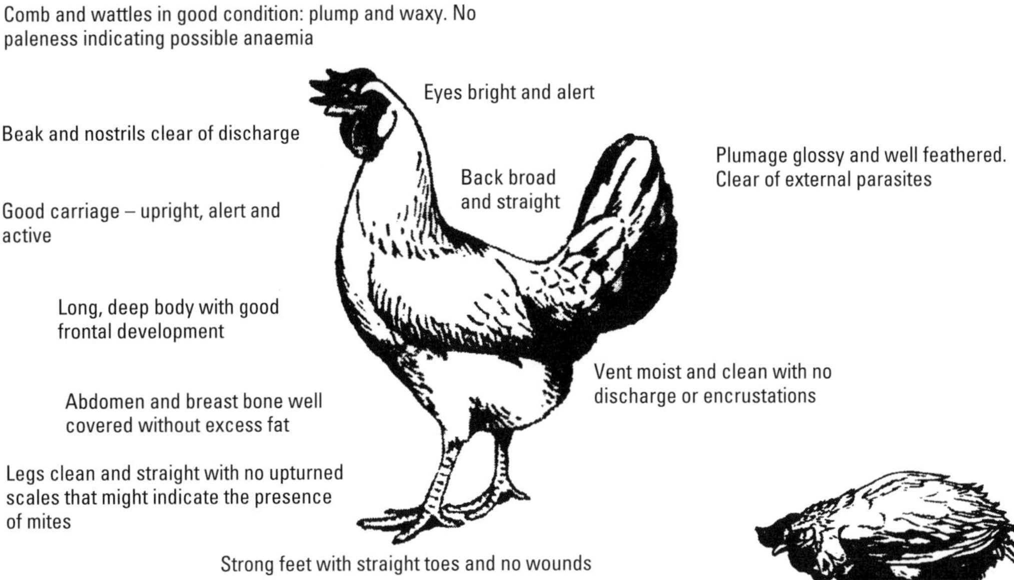

Comb and wattles in good condition: plump and waxy. No paleness indicating possible anaemia

Eyes bright and alert

Beak and nostrils clear of discharge

Plumage glossy and well feathered. Clear of external parasites

Back broad and straight

Good carriage – upright, alert and active

Long, deep body with good frontal development

Vent moist and clean with no discharge or encrustations

Abdomen and breast bone well covered without excess fat

Legs clean and straight with no upturned scales that might indicate the presence of mites

Strong feet with straight toes and no wounds

What to look for in a healthy chicken.

Typical stance of a sick chicken.

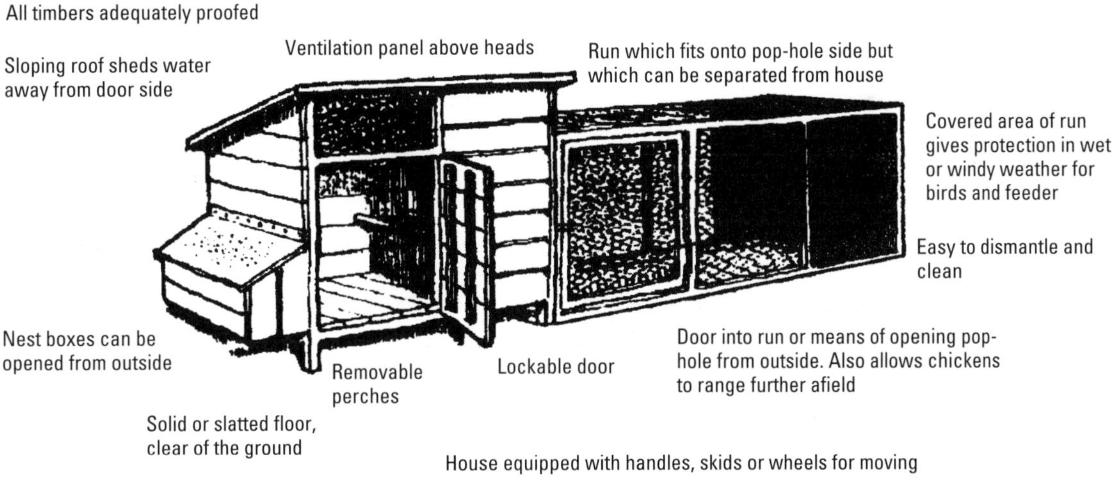

All timbers adequately proofed

Ventilation panel above heads

Run which fits onto pop-hole side but which can be separated from house

Sloping roof sheds water away from door side

Covered area of run gives protection in wet or windy weather for birds and feeder

Easy to dismantle and clean

Nest boxes can be opened from outside

Door into run or means of opening pop-hole from outside. Also allows chickens to range further afield

Removable perches

Lockable door

Solid or slatted floor, clear of the ground

House equipped with handles, skids or wheels for moving

Features of a house and run.

Materials

The most commonly used material is wood, and this has the advantage of being relatively cheap, strong and warm. The disadvantage is that it needs to be treated against damp, otherwise it will rot. The cheapest small houses come untreated, and must be treated with appropriate proofing. Buy one that is non-toxic to birds and animals. Most small houses, however, come ready proofed, and those that have been pressure-proofed will last longest. The walls of timber houses may be of exterior grade plywood, match boarding or overlapping weatherboarding. The latter is the strongest, but also the heaviest, a factor to be borne in mind for a house that needs to be moved.

For larger numbers, such as a small, free-range, commercial laying or table-bird flock, polyester-coated pressed steel lined with internal insulation is increasingly being used. Some houses are a combination of steel and wood, the advantages being that they can be easily dismantled for moving, are less likely to harbour mites, and are easier to keep clean. The disadvantages are that they are not generally available for fewer than 100 birds; they are comparatively expensive; and ventilation needs to be carefully controlled in order to avoid condensation.

In small houses, roofing may be of Onduline corrugated bitumen, boarding, or wood covered with bitumen felt – though the latter is more likely to harbour mites than the others. Larger houses may have Onduline, polyester-coated pressed steel, or aluminium sheeting; these are insulated on the inside.

Poultry houses are much easier to move if they are equipped with wheels.

Mobility
There is little doubt that mobile houses for small flocks provide the best environment, but how they are to be moved is an important consideration. Depending on the size, there may be carrying handles, skids or wheels. The latter are undoubtedly the best for small domestic houses, and many suppliers now sell them as an optional extra for their range of houses. Some houses can be completely dismantled for moving, while others can be pulled along by tractor.

Temperature
The optimum temperature for a chicken is 21°C, a reminder of its jungle origins. If there is sufficient insulation on the roof and walls of the house to prevent excessive heat loss, the birds themselves

A small house and run for a trio of Speckledy hens in a small kitchen garden.

will generate enough heat. Recommended roof insulation is to a value of U=0.5W/m/°C.

It is important that the house is not too big for the number of birds, otherwise in really cold weather they will not be able to keep each other warm. Putting a small maximum/minimum thermometer inside the house, out of pecking reach, is a good idea. Warmth always needs to be balanced against ventilation, which is crucial if lung infections are to be avoided in the flock.

Ventilation
Ridge ventilation is common in both large and small houses, where a ridge across the top of the roof provides an open space, allowing stale air to escape but preventing rain from coming in. Within the house there should not be a detectable smell of ammonia or inhalable dust that exceeds 10mg per volumetric metre.

Ridge ventilation is used in association with side vents that can be opened or closed as required. In a very small house, the vents may be in the form of holes with mesh across them; galvanized, welded mesh is better than poultry netting because it is stronger and less likely to rust. In some houses there may be windows that act as side vents. Polycarbonate is a better option than glass because it is warmer and less prone to breaking.

Humidity
Inside the house this should ideally be between 60–65 per cent, and it is a good indication of how effective the ventilation is. Small humidity gauges are cheap to buy and can easily be installed inside the

house, out of pecking range. If the humidity is over 75 per cent, then ventilation should be increased.

Floors

They may be solid, slatted, or of rigid wire mesh. Solid floors are warmer but get very dirty, where slatted ones allow droppings to fall through. The gaps between slats should not be more than 2.5cm (1in) wide, otherwise rats may gain entry – though bear in mind that mice can still get in through a gap this size. Whatever type of floor it has, the house should ideally be raised at least 30cm (1ft) above the ground as this prevents predators taking up residence underneath – and if they do, cats and terriers can get in to clear them out.

With a small house, solid floors can be lined with a thick polythene sheet with wood shavings, chopped straw or sawdust on top; this makes it easy to keep the house clean, because the whole thing can be removed and replenished as required. Alternatively, a droppings board that can be pulled out is an effective and quick way of clearing the floor area. In a static house, the area beneath a slatted floor may be a droppings pit, in which case it is normal for the pit to be emptied less frequently. The pit will need a metal grille to prevent the birds gaining access to the droppings from outside.

Perches

The chicken is a perching bird, and always prefers to roost at night. Ideally, perches should be slightly rounded at the top, and 4–5cm (1½–2in) wide; large birds such as Brahmas and Cochins may need slightly wider ones.

The perches may be at the same level or in staggered rows, but the bottom one should not be more than 60cm (2ft) from the ground if heavy birds are kept: much higher than this and they will risk landing heavily and suffering foot injuries. No less than 18cm (7.5in) perch space should be allowed for average-sized birds, while 20cm (8in) is better for larger breeds (a commercial hybrid layer is considered to be an average-sized bird). Perches should be easy to remove for cleaning.

Pop-holes

Pop-holes provide the entrance or exit points for the birds. For average-sized chickens they are normally around 30cm high × 25cm wide (12in × 10in); for large birds a better size is 38cm × 30cm (15in × 12in). In commercial, free-range houses, pop-holes are made bigger in order to encourage the birds out and also to let several birds out at the same time. The minimum dimensions for these are 45cm high × 2m wide (18in × 79in).

Pop-holes may have doors that slide up or pull down to form a ramp for ease of exit. If the house is inside a run, it is an advantage to be able to open the pop-hole from outside the run. If the pop-hole is above ground but does not have a pull-down door that forms a ramp, a separate ramp will be needed. A ladder type is suitable, or a solid construction as long as it has strips nailed on for ease of access. Larger houses will need to have wide ramps, in relation to the number of birds.

Nest Boxes

In small domestic houses there should be one nest box for every three hens; in commercial houses, the recommended minimum ratio is one nest box per six birds. The average size is 30cm high × 30cm deep × 25cm wide (12in × 12in × 10in), but again, the dimensions should be greater for large birds. If the nest boxes are off the ground, an alighting perch should be provided to encourage and facilitate access.

Nesting boxes can be lined with sawdust, wood shavings, chopped straw or a plastic substitute such as Astroturf. Wood shavings may mark brown,

A free-ranging flock of Bovans Goldline in a small, movable house on wheels. They are protected by electric fencing, and the trees provide shade, wind protection and a sense of security.

A flock of Cobb table birds being reared organically. The movable house has extra wide pop-holes, and the area is protected by electric netting. Photo: Associated Poultry

speckled eggs when they are newly laid and damp, and owners of these breeds may wish to consider an alternative.

Runs

Chickens need access to the outside where they can forage and take dust baths; the area may be an attached run, part of a garden, an orchard or a field. If buying a run, order it from the same supplier as the house so that it fits properly. Some suppliers have run extensions if you subsequently wish to enlarge the area. If predators are a problem, garden netting can be placed over the run.

Small commercial houses will generally be in an area of grassland that is surrounded by electric poultry fencing. Note that maximum flock density requirements outside are now standardized throughout the EU. The grazing area needs to be changed frequently in order to avoid damage to the grass, and to reduce the incidence of pests. Again, the most effective way of controlling access to particular areas of pasture is to use electric poultry netting. A good rule of thumb is to move the house or birds as soon as the grass shows evidence of wear and tear, though the timing of this will obviously vary depending on the size of the house and the number of birds; thus a small, garden-scale house with attached run may need to be moved every day. It is not a good idea to move a combined house and run with the chickens in the run, in case their legs are damaged.

Once an area of pasture has been vacated by poultry, it can be raked and limed (harrowed on a field scale) in order to disperse the droppings; the rain will wash the residues into the soil. Alternatively grow an acid-tolerant crop such as oats or potatoes on the vacated area, then apply lime, followed by a sowing of new grass ley or wheat.

Sometimes a small flock may be given a special winter run if there is a shortage of clean grass for them. In this, a layer of wood chippings or pea gravel 15cm (6in) or so deep can be used to absorb any water within the run and prevent a mud bath; or a concrete run can be put in, and this can be washed down. A large, shallow box of sand or fine soil can be given to the birds for their scratching activities at this time – though it may be more appropriate to house the birds in a garden shed or barn until the weather improves. We utilized a stable as winter housing. From here, the chickens had access to the orchard via an open-air yard. In really bad weather or in heavy snowfall they were allowed access to the yard only.

Breeds of Chicken

A large selection of breeds is available. Pure breeds will produce progeny identical to the parents when a male and female of that breed are mated; crosses are where two different breeds are mated. With some crosses it is possible to identify newly hatched chicks as males or females by differences in feather colour; these are referred to as sex-linked crosses.

Breeds are also divided into light and heavy types: the light ones are best for egg production, while the heavy breeds are more appropriate for the table. The latter are also called 'sitting breeds' because they have a greater tendency to become broody (when they sit on, and incubate, a clutch of eggs). Some breeds are called 'dual purpose' because they lay a reasonable number of eggs, and are also heavy enough to be used for the table.

Other terms used in relation to poultry are 'hard feathered' and 'soft feathered'. 'Hard-feathered' birds have tight, very close-fitting plumage – for example, game birds such as Old English Game and Indian Game. All other breeds are 'soft feathered' in that the plumage is generally looser and fluffier. The Silkie breed is distinctive because, unlike all other soft-feather breeds, its feathers do not have barbs, so the plumage is more like soft fur than feathers.

Bantam breeds are naturally-occurring small chickens. There are also small chickens that have been bred down, as miniature versions of large breeds; these are popular with show breeders and in domestic flocks where space may be limited.

Finally, utility breeds are those that have been bred for their good production of eggs or table

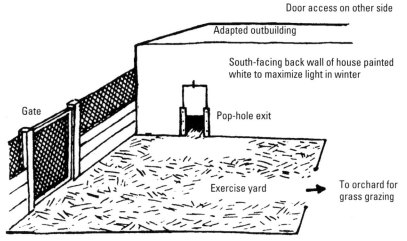

Initially, chopped straw was used in the exercise yard, but in winter when conditions were muddy, it was replaced with a thick layer of wood chips. This allows water to drain through.

Inside the house

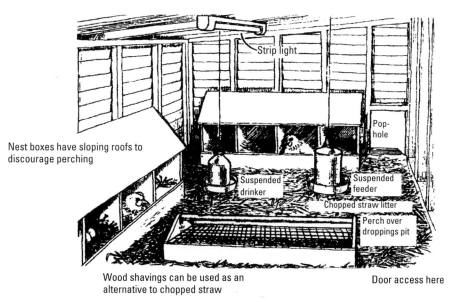

The author's system of housing laying hens.

Flock Density on Free Range	
Inside the house	7 birds per 1sq m (10sq ft)
Free-range and organic layers	1,000 birds per hectare (400 per acre) 1 bird per 10sq m (100sq ft)
Free-range and organic table birds	2,500 birds per hectare (1,000 per acre) 1 bird per 1sq m (10sq ft)

qualities, while 'show' or 'fancy' breeds are more appropriate for exhibition and may not be particularly productive. After World War II, the trend was to hybridize different strains in order to produce birds suitable for intensive production, and the old breeds were gradually usurped as commercial varieties; indeed, some were kept going only by the dedication of small breeders. Inevitably they tended to be bred for showing, and so gradually lost their productive capacity. In recent years, however, efforts are being made to improve their performance again, by selective breeding. If the choice is to be a pure breed, it is therefore worth trying to find a good utility strain, rather than merely relying on a named breed. The emphasis in this book is on utility breeds.

Traditional Breeds

Rhode Island Red: Bred in America and introduced to Britain in 1904, this is one of the traditional heavy breeds that is also regarded as a dual-purpose bird for eggs or the table. The male was frequently crossed with Light Sussex or Barred Plymouth Rock females to produce a commercial sex-link cross. With these crosses, the practice was to use the females as replacement layers while the males were raised for the table.

The Rhode Island Red has had a major influence on the development of brown-egg-laying hybrids. In its heyday it was capable of producing up to 300 eggs, though both productivity and fertility declined as the emphasis shifted to exhibition requirements.

Light Sussex: A heavy British breed, the Light Sussex was also regarded as dual purpose, and for many years was Britain's premier table bird as well as being a prolific layer of tinted brown eggs; the appearance of hybrids eventually sent it into decline. Its propensity for staying broody and sitting well made it popular for incubating and raising chicks, before artificial incubation became widespread.

White Leghorn: The Mediterranean Leghorn is a light breed that has had a major influence on egg production, with the white variety being particularly prolific. Most hybrid layers of white eggs have White Leghorn in their genetic development.

Barred Plymouth Rock: A traditional, heavy American breed, the Plymouth Rock was also one of the most important dual-purpose breeds, providing eggs and table birds.

A breeding flock of Light Sussex at the Domestic Fowl Trust. The house has skids so that it can be towed when moving is necessary.

A flock of utility Barred Plymouth rocks.

Maran: An old, heavy French breed, the Maran is best known for its dark brown, speckled eggs, although its productivity has declined. It was also a traditional table breed.

Welsummer: A light Dutch breed, the Welsummer was introduced to Britain in the 1920s and quickly became popular; it produces brown eggs, with a matt shell rather than a glossy one.

Wyandotte: This heavy breed was developed in the USA for both its egg-laying and its table qualities. It was named after a native American tribe and lays tinted eggs.

Commercial Free-Range Egg Breeds
Hybrids are normally the choice of the commercial producer because they have been developed for maximum production from several different breeds and strains. A strain is a family group that has been selected and bred for particular qualities, such as egg numbers and quality. Originally, hybrids were developed for intensive enterprises; however, more recently breeders have responded to the growing demands of the free-range sector by producing birds that are more suited to outside conditions. This entails having a heavier bodyweight and tighter feathering, and an emphasis on docility, while maintaining good egg numbers and quality. The heavier bodyweight does mean that outdoor hybrids are slightly later in maturing than are the lighter cage hybrids; they also consume more food. It is interesting that many of these new strains are given names rather than numbers, which is usually the case with hybrids; perhaps the reason for this is that consumers can relate more readily to concepts such as 'free-range' and 'traditional produce' if eggs come from a breed with a name.

The following hybridized breeds are those that have been developed for the free-range egg sector in recent years. They are suitable for commercial and domestic situations, but as they are crosses, they will not necessarily breed true.

Black Sex Link: Based on a cross between the Rhode Island Red male and the Barred Plymouth Rock hen, this hybrid is so named because it has black plumage and the sex of the chicks can be distinguished at just a day old. Like all the birds bred for free-range, it matures slightly later than cage hybrids, and has a higher bodyweight. Different strains are available under different names, depending on the breeders. In Britain, the popular Black Rock is bred for its tight feathering; it

produces brown eggs. The Hebden Black is also reared in Britain, while the Bovans Nera is bred in the Netherlands.

Cream Legbar: Green and blue-green eggs have become popular for speciality markets. The Cream Legbar was developed from the Leghorn and Araucana breeds, the latter providing the genetic factor for the eggshell colour, as well as a head crest. Although egg numbers are not large, producers like to have some of these birds so they can provide 'multi-colour' egg packs, or for special occasions such as Halloween. The Old Cotswold Legbar is a commercial strain of the Cream Legbar.

Hisex Ranger: This hybrid was bred specifically for the free-range sector; it first appeared in 1994, and is based on existing, high-yielding Hisex female lines but with a different male line. The plumage is light brown with white tail feathers, and the eggs are brown.

Lohmann Brown: Bred in Germany, this is a hybrid based on Rhode Island Red-derived hybrids for the free-range sector. It has brown plumage with whitish tail feathers and attractive gold circling around the neck. It lays medium brown eggs. A more recent development has been the breeding of the Lohmann Tradition, a layer of large eggs for speciality markets.

Speckledy: Based on the Maran and other traditional strains originally held by Cyril Thornber in Britain, the Speckledy has grey and white barred plumage. It produces medium to dark brown eggs with attractive speckling.

Columbian Blacktail: Based on the Rhode Island and bred for docility as well as laying ability, this free-range hybrid has golden brown plumage and distinctive black tail feathers. The eggs are mid-brown. It is also called the Calder Ranger.

Babcock B380: Another layer of brown eggs, developed for the free-range sector. It is capable of producing up to 307 eggs in up to 72 weeks.

Hybrid layers: Mention should also be made of other hybrids that, although they have not been specifically bred for free-range conditions, are still capable of high yields in outside conditions as long as they are not in particularly exposed areas. They include the brown-feathered ISA Brown (originally called the Warren), the Hisex Brown, Shaver 579, Bovans Goldline and Hyline.

Taking dustbaths is an instinctive pattern of behaviour, as this Lohmann Brown hen is demonstrating.

For white-shelled eggs there is nothing to beat a commercial strain of White Leghorn: production is over 300 eggs, and several strains are available from breeding companies. The Bovans White, for example, is a strain developed in the Netherlands, but also available in Britain under the name White Star.

Table Breeds

Traditional heavy breeds that were used for the table included the Light Sussex, the Plymouth Rock and the Cornish, the latter often referred to as 'Indian Game'. But from 1953 onwards, when food rationing ceased in the UK, a large-scale factory mentality began to appear, in response to the Americans who had been developing fast-growing hybrids specifically for the table. Known as 'broilers', from the American term 'broil' ('grill' in Britain), they soon became dominant, and commercial pure-breed flocks went into a decline. The problem with most pure breeds now is that they do not produce enough plump breast meat, and are often long-legged and lacking in docility. Organizations such as the Utility Breeders' Association and the Rare Breeds Survival Trust's Poultry Project are trying to reverse the situation, by encouraging breeders to develop more productive strains through selective breeding.

White-feathered breeds: In recent years the choice of table birds has been largely confined to white-feathered birds such as the Cobb and Ross broilers. However, the fact that they were developed for the intensive sector does not mean that they cannot adapt to outside conditions. It is true that under intensive conditions some grow so quickly that they develop leg problems; but if fed less intensively on a predominantly grain diet, and allowed outside exercise, growth is slower and problems are less likely. We never had problems raising Cobbs on free range – though admittedly they were all hens, and some free-range producers have reported problems with the heavier males.

Recent developments have produced white-feathered hybrids such as the Hybro or Sherwood White that are intrinsically slower growing, and specifically aimed at the free-range sector.

Coloured-feather breeds: When consumer reaction against intensive farming methods began, there was an associated interest in using traditional breeds again. Unfortunately, most of the pure breeds were no longer regarded as being commercially viable. Birds were therefore needed that looked like traditional breeds, but were more prolific, and produced ample breast meat without being too rangy. It was the French who met the challenge, and developed coloured hybrids as table birds for their *Label Rouge* (Red Label) designation, a description introduced to describe slow-growing birds raised on free range.

One of the main breeding companies involved in breeding stock for *Label Rouge* production is SASSO. They have developed primary breeders that can be used to produce a wide range of different coloured table birds to suit different markets: these include red, grey, black and other colour variations. Areas of France have traditionally had their own local type and colour of table bird. The principle is that small, prolific hybrid hens are crossed with pure-bred terminal sires. The females have a recessive gene so that the offspring always resemble the father, and therefore the traditional qualities associated with traditional breeds; but they also have the plumpness and prolific characteristics associated with more modern strains.

Hubbard-ISA (France) are also involved in the production of red and other coloured-feather table birds, such as their primary breeders Redbro and JA57 females. As with the SASSO breeds, the birds are slow growers without the leg problems sometimes associated with intensive broilers, and are bred for good breast conformation and meat texture.

Rearing companies who are supplying the day-old chicks often use their own brand names: Poulet Bronze, for example, are Hubbard-ISA Redbros from a hatchery in Lincolnshire, which supplies the organic sector. Poulet Gaulois is the name of the Sasso 551 that is imported into Britain by a company in Norwich.

Redbro table chickens being reared organically. The house has suspended feeders and drinkers, and wide pop-holes for ease of exit. Photo: Hubbard-ISA

Showing poultry is popular with many and helps to maintain breed standards. This trio of Buff Pekin bantams is on display at the National Poultry Show.

In the USA, organic and free-range production lags behind Europe, and there are fewer breeds developed for extensive conditions. Exceptions are hybrid-cross broilers based on New Hampshire Reds to produce red-feathered birds, and black-feathered broilers based on Black Australorps.

Breeds for Showing and Selling

Breeds that are kept for showing are required to be traditional and standardized so they will breed true. Each breed has its own club that is responsible for drawing up a 'standard of excellence', against which specimens of the breed can be measured. In Britain, the breed clubs are overseen by the Poultry Club of Great Britain, while the USA has the American Poultry Association; however, there are variations in their relative standards. The panel (*see* next page) shows many of the breeds that are currently available, although some are rare and localized.

Acquiring Birds

The most convenient time to buy birds is when they are coming up to point-of-lay. If bought at around sixteen weeks, this gives them the chance to settle down in their new surroundings before laying starts at around twenty-one weeks. Early laying is not a good idea for free-ranging birds, because they need to develop a good physical framework to cope with the demands of being outside as well as of laying. They need to be confined to their house for twenty-four hours (with food and water) and then let out: this allows 'home' to be imprinted so that they go back to the house to lay eggs and to perch at night.

Birds can be bought as day-olds, but they will need protected conditions until they are hardy, at around six weeks. It is also possible to acquire them at this age or later, when mortalities are less likely.

New birds should not be mixed with old birds, not only to reduce the chances of infection, but also to prevent fighting as a result of the pecking order. If there is no choice but to mix different ages, keep the new ones confined in a separate house, and run them next to the original flock so they can see each other. After about a week they will be used to each other, especially if their grain ration has been placed in their respective runs, on either side of the dividing wire. This procedure is also a good one to use where any bird has had to be temporarily separated from the rest of the flock, so that when reintroduced, she is not attacked as a newcomer.

Feeding

The digestive system of the chicken is fairly small so it is important to provide food that contains a balance of the necessary nutrients. Feeding too much of one type of food at the expense of another may result in a nutritional imbalance and digestive upset.

Compound feeds will cater for all a chicken's nutritional requirements. They are available as pellets or as dry meal (mash) and provide all the necessary proteins, carbohydrates, oils, minerals and vitamins. Proprietary free-range rations specify on the label the amount of protein in the feed: this is usually 16 per cent, 17 per cent or 18 per cent, and the lower the percentage, the more necessary it is to have good pasture from where the birds can obtain a proportion of their protein requirements. There are also starter rations or chick crumbs for young birds up to the age of five weeks; these have a protein

Table of Breeds

(Key: E = egg breed; T = table breed; S = show breed
B = brown eggs; W = white eggs; Tn = tinted eggs; G = green eggs)

Heavy Breeds

Australorp (E, T, B)	Frizzle (S, W)	Norfolk Grey (S, Tn)
Barnevelder (E, T, B)	German Langshan (S, B)	North Holland Blue (S, Tn)
Brahma (S, B)	Houdan (S, W)	Orloff (S, Tn)
Bresse (T)	Ixworth (T, Tn)	Orpington (S, B)
Cochin (S, Tn)	Jersey Giant (T, S)	Plymouth Rock (T, E, B)
Crévecoeur (S, W)	La Flèche (S, Tn)	Rhodebar
Croad Langshan (S, W)	Marans (T, E, B)	Rhode Island Red (T, E, B)
Dominique (E, B)	Modern Langshan	Sussex (T, E, Tn)
Dorking (S, Tn)	Naked Neck (T, Tn)	Wyandotte (T, E, Tn)
Faverolles (S,)	New Hampshire Red (T, E, B)	Wybar

Light Breeds

Ancona (S, W)	Kraienköppe (S, W)	Sicilian Buttercup (S, W)
Andalusian (S, W)	Lakenvelder (S, Tn)	Silkie (S, W)
Appenzeller (S, W)	Legbar (E, W)	Spanish (S, W)
Araucana (S, G)	Leghorn (E, W)	Sulmtaler (S, B)
Augsburger	Marsh Daisy (S, Tn)	Sultan (S, W)
Brabanter	Minorca (S, W)	Sumatra (S, W)
Brakel (S, W)	Old English Pheasant Fowl (S, W)	Uilebaard
Breda (S, W)	Poland (S, W)	Vorverk (S, Tn)
Campine (S, W)	Redcap (S, W)	Welbar (S,)
Fayoumi (S, W)	Rhienlander	Welsummer (E, B)
Friesian (S, W)	Rumpless Araucana (G)	Yokohama (S)
Hamburgh	Scots Dumpy (S, W)	
Italiener	Scots Grey (S, W)	

True Bantams

Belgian Barbu D'Anvers (S)	Belgian Barbu de Watermael (S)	Nankin (S, E, Tn)
Belgian Rumpless D'Anvers (S)	Booted (S)	Pekin (S, W)
Belgian Barbu D'Uccle (S)	Dutch (S, E, Tn)	Rosecomb (S, W)
Belgian Rumpless D'Uccle (S)	Japanese (S, W)	Sebright (S)
		Tuzo (S, Tn)

Hard-Feathered (Game Birds)

Asil (S, T)	Malay (S, Tn)	Rumpless Game (S, Tn)
Belgian Game (S)	Modern Game (S, Tn)	Shamo (S, W)
Carlisle Old English Game (S)	Nankin Shamo (S)	Tuzo (S)
Indian Game (Cornish) (S, T, Tn)	Oxford Old English Game (S)	Yamato-Gunkei (S)
Ko-Shamo (S)		

Note: Any of these breeds may be regarded as show breeds, but only those that have no potential as utility breeds are indicated as such with the letter 'S'. The egg colour specified is a general guide, because there is considerable variation: for instance, brown may be light, mid-brown or dark, white may include cream, while tinted includes pinkish-brown. There are bantam versions of many of the large fowl.

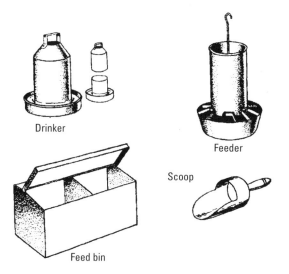

Drinker

Feeder

Scoop

Feed bin

Feeding equipment.

Free-range layer's pellets being put in a plastic feeder for a small flock of chickens. The feeder can be suspended if necessary.

level of around 18.5 per cent. Grower rations are fed from five to eighteen weeks, and have a protein level of about 15 per cent. After this, the free-range layer's ration referred to above can be given.

Free-range compound feeds do not have the range of additives contained in those feeds destined for the intensive sector; these include artificial colouring agents to make the yolks bright yellow and, for broilers, low levels of antibiotics or growth promoters. Free-range birds obtain a certain level of natural yolk colouring from grass, although the yolks may be paler in winter when the grass is not growing. Grassmeal and alfafa (lucerne) are often used in compound feeds. Free-range table birds grow more slowly and naturally without growth promoters. The best choice of compound feed is an organic ration, although it is the most expensive, since it does not allow artificial additives or genetically modified ingredients. The next best is a normal free-range ration. Check the label for additives!

Whole grain is popular with chickens, and many poultry keepers feed a compound ration in the morning, with grain in the afternoon. Chick corn or chopped grain is also available from many suppliers. If grain is put on the ground, it satisfies the hens' instinctive need to scratch, as well as encouraging them to range all over the site. Wheat is the best whole grain to give, in association with a compound ration. As a rough guide, one hybrid hen will take 130g (5oz) of compound feed and 15g (0.5oz) of grain a day, though heavier birds will need more. In winter, the grain ration should be increased to cater for the extra demands made on the system. If a 'grain only' ration is to be fed, it should be a mixture of grains. Watch out for the occasional greedy hen that picks out the soya and maize from a mixed grain ration, leaving the other grains behind: this can lead to an over-fat bird.

Insoluble grit is needed so that the gizzard is able to break down grains in the digestive system. Free-ranging birds will usually pick up small stones, but it is a good idea to make a container of poultry grit available so that they can help themselves as required. Crushed oystershell can be mixed in with the grit, to ensure that the birds are getting enough calcium for strong egg shells. Fresh, cold water is essential at all times. Five adult hybrids drink an average of 1 litre (1.8 pints) a day. In hot weather or where large birds are kept, this can double, or even treble.

Breeding

A cockerel is not required for an egg-laying flock and should not be allowed to run with the hens if eggs are to be sold for eating; it is a myth that hens will not lay well unless there is cockerel, besides which his crowing is likely to disturb close neighbours. A male is only required for specific breeding purposes, and either he should have his own house and pen to which hens are introduced as necessary, or he should run with a breeding flock (as distinct from a laying flock). His spurs should be kept trimmed so that he does not damage the hens' sides: soften them with oil for a few days before trimming, and cut the ends only. Repeat this over a period of time and they will gradually be reduced to a less damaging size.

If hens have been running with other males from which you do not wish to breed, no eggs for hatching should be taken for at least three weeks, and preferably a month, after they have been introduced to the breeding male; this is because sperms can remain viable for three weeks in the hen's oviduct. Breeding birds should carry some form of identification, such as leg rings, so that breeding records can be kept. A record of any medication administered is also necessary for commercial birds.

Although all birds will do well on a layer's ration, the ideal is to feed a compound breeder's ration to breeding birds. This has a protein level of about 17 per cent, and a higher level of vitamins and minerals, both of which will enhance fertility and minimize the chances of chicks with deformities.

Breeding birds should be healthy, productive and good examples of their type. If a hen is broody, she can incubate her own eggs, or those of other birds. She needs a protected nesting area, such as a small ark and run, and should be checked to ensure she is free of lice and mites. She will emerge from the nest about once a day to pass droppings and to feed, so a feeder and drinker should be placed close by.

Eggs that are required for incubation can be taken and stored for a week; after that the hatchability declines. The ideal storage conditions are a temperature range of 15–18°C (60–65°F), and a relative humidity of 75 per cent. Eggs to be incubated should be washed in warm water, to which an egg sanitant has been added: the temperature needs to be higher than that of the shell, otherwise dirt and pathogens may be drawn in through the shell pores (comfortably warm to the hand is about right). The eggs should be immersed in the solution for a few seconds so that mycoplasma organisms and other pathogens are killed. There is a far greater chance of hatching healthy chicks if this procedure is followed.

Modern incubators are preferable to old models, where the control systems can be very imprecise;

these days even a small table-top incubator is usually equipped with an electronic thermostat and automatic egg-turning facilities. Eggs need to be turned several times a day, every day, for the first eighteen days of the twenty-one-day incubation period. This is tiresome if it needs to be done manually.

The incubator should be thoroughly cleaned and disinfected before use, and run for at least twenty-four hours before eggs are introduced. It is much better to place the incubator in a warm room than in a cold outhouse, where temperatures and humidity can fluctuate wildly. It is also essential to follow the manufacturer's instructions, for they will vary, depending on the model.

The incubation period lasts an average of twenty-one days, though there may be earlier or later hatchings. Around day nineteen, the chick will begin to break a hole in the shell, a process called 'pipping'. At this time, turning of the eggs should cease, while the temperature is slightly reduced and the relative humidity increased, as shown in the box on page 129. Commercially, separate incubators and hatchers are used.

After a week in the incubator eggs can be 'candled': this entails holding them against a bright light to check on progress. If there is a developing embryo, it will be seen as a blob with blood vessels radiating outwards like a starfish. Infertile eggs can be discarded. Candling can also be used to check on the humidity, as well as checking on egg quality generally, where eggs are being sold.

Once the chicks have hatched, they will need protected brooder conditions with a heat lamp to keep them warm. Electrical or gas-powered lamps are available, and these should be suspended low enough for the chicks to move about comfortably beneath them. Gauge if you have them at the right height by the way the chicks behave: thus if they are all clustered in the middle they are probably cold, so the lamp needs to be lowered; and if spread around the edges with none in the centre, they are too hot, so it needs to be raised. In our experience this is a far more accurate guide than worrying about providing exact temperatures.

If there are any chicks that are slow to learn to feed, drop some crumbs of feed onto a sheet of paper to attract their attention. This is what the mother hen does, and they will instinctively go and investigate. Dipping their beaks in water will also teach them to drink.

Once the birds are fully feathered, the lamp will no longer be required, and they are hardy enough to go out. In fact, the earlier they go out, the better, for this helps to improve muscle tone. If the chicks have

Home-Made Layer's Feed

	%
Wheat	60
Grassmeal	15
Full fat soya	15
Fishmeal	5
Molasses	2.5
Vitamin and mineral supplement	1.5
Di-calcium phosphate	1.0

Note: The fishmeal can be replaced with extra soya, maize or peas.

Nutrient Requirements and Sources

The necessary nutrients are indicated as well as some of their sources. These may be of interest to small flock owners wishing to use a certain proportion of garden plants and produce for supplementary feeding.

Nutrient	Sources
Carbohydrates	Wheat, barley, oats, maize, millet, potatoes (cooked)
Protein	Soya, peas, beans, maize, sunflowers, barley, oats, wheat, grass, clover, lucerne, dried yeast
Lysine	Soya
Methionine	Soya
Tryptophan	Soya
Fats (Oils)	Sunflowers, linseed and other vegetable oils
Phosphorus	Lucerne, grassmeal, oats, dandelions, dried yeast
Calcium	Oystershell, calcified seaweed, ground limestone, lucerne, molasses, dried yeast, green vegetables
Sodium	Maize, sunflowers, lucerne, molasses, dried yeast
Chloride	Molasses, lucerne, grassmeal
Linoleic acid	Maize, sunflowers, vegetable oils
Vitamin A	Grass and grassmeal, maize, kale and brassicas, carrots, nettles
Vitamin B_1 (Thiamin)	Most cereals
Vitamin B_2 (Riboflavin)	Grass and grassmeal, dried yeast, soya
Vitamin B_3 (Niacin)	Wheat and other cereals, bread
Vitamin B_5 (Pantothenic acid)	Dried yeast, molasses, grassmeal, comfrey
Vitamin B_6 (Pyridoxine)	Soya, wheat, dried yeast
Vitamin B_{12} (Cobalmin)	Comfrey, calcified seaweed
Vitamin D_3	Sunshine, wheat, maize and other cereals
Vitamin E	Maize, wheat, grass, grassmeal, kale and brassicas, lettuce
Vitamin H (Biotin)	Maize, yeast, lucerne, grass and grassmeal
Vitamin K	Lucerne, grass and grassmeal, kale and brassicas
Nicotinic acid	Sunflowers, dried yeast, wheat, maize
Choline	Soya, maize
Folic acid	Wheat, grass, chicory, chickweed, parsley, dandelions, nettles
Manganese	Dried yeast, molasses, lucerne, wheat, maize, millet, oats
Iron	Nettles, parsley, chicory, chickweed, dandelions
Zinc	Dried yeast, lucerne, molasses, maize, wheat, sunflowers, soya
Copper	Cereals, beans, soil (if free-ranging, and copper is present in soil in that area)
Potassium	Molasses, lucerne, maize, wheat, soya, sunflowers, potato, dried yeast
Cobalt	Calcified seaweed
Selenium	Green vegetables, dried yeast, maize, lucerne
Magnesium	Oats, beans, soya, grass and green foods, spinach, Good King Henry
Iodine	Washed seaweeds, calcified seaweed
Yolk enhancers	Grass and grassmeal, maize, green vegetables, pot marigold petals

Note: If a basic diet of compound feed is given, all the necessary ingredients will be included. If the basic ration is grain only, give a range of grains and ensure that minerals and vitamins are provided. Fresh green foods and other garden produce can be suspended for the birds to peck if they do not have extensive grazing. Do not give them stale or stringy foods as these can cause digestive upsets or a blockage in the crop.

Eggs should be washed in a warm solution of egg sanitant before being put in the incubator. In this way, losses to disease are minimized.

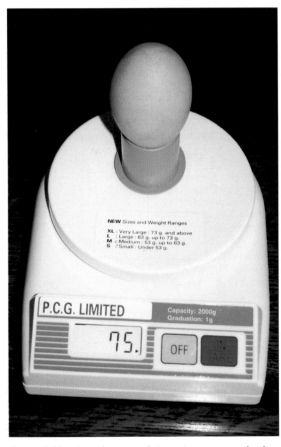

Egg weighing in order to grade into sizes, or as a check before incubation.

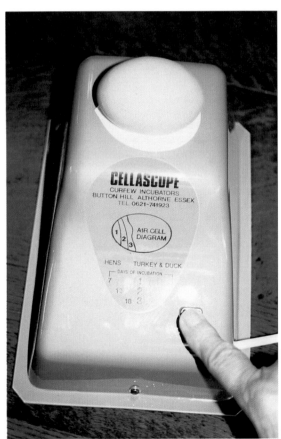

A candler for checking the contents of an egg.

Chicks hatching in an incubator. They do not need food for twenty-four hours because the remnants of the yolk are still in the abdomen.

been incubated and hatched by a broody hen, they will be looked after by her for five to six weeks, after which time she will lose interest in them.

For the first few weeks until they are feathered, an 18 per cent protein starter feed of chick crumbs can be given. From the age of five weeks, the chicks can gradually go over to a 15 per cent protein ration of grower pellets; alternatively, a mixed ration of chopped grains can be given. Once they are at point-of-lay (POL), their main ration will be a layer's ration in the morning, with wheat in the afternoon, as referred to earlier. All these rations are available as additive-free feeds.

Table Birds

Day-old table birds need the same protected conditions until they are fully feathered and ready to go outside. Initially, they can be fed on chick crumbs. There is an organic version that does not contain an anti-coccidiostat, although Amprolium can be used if there is a problem. As the chicks grow, they can be switched to a table-poultry feed; for instance maize or 'corn-fed' chickens develop a skin with a golden hue. Food and water must be available on an *ad lib* basis at all times. Several companies now supply organic cereal mixes without genetically modified ingredients or growth promoters. Producers who are selling organically produced table birds or eggs must be registered with an approved certification body.

A movable house, placed where there is tree shade and protection, is ideal for table birds, while they will benefit from the exercise and grazing provided by a grass paddock. Normal free-range birds will be slaughtered from fifty-six days onwards, but organic ones will not be killed until they are at least eighty-one days. Typical live weights for organic production at eighty-one days are 2–2.5kg (4.5–5.5lb), with a 3:1 feed conversion ratio. (Feed conversion ratio, or FCR, is the amount of weight gained in relation to the food eaten. For example, if 1.35kg (3lb) of food is eaten for every 450g (1lb) weight gain, the FCR would be 3:1.)

Commercial producers either slaughter the birds on site, or send them to the nearest licensed slaughterhouse for processing. If it takes place on the smallholding, planning permission may be

Optimum Conditions for Incubating and Hatching Chickens		
Room where incubator is housed:	24°C (76°F)	Relative humidity: 60%
Inside the incubator:	day 1 – 18: 37.5°C (100°F)	Relative humidity: 52%
	day 19 – 21: 37°C (99°F)	Relative humidity: 75%

Newly hatched chicks under a heat lamp.

A broody hen and chicks in a small broody coop and run.

required from the local authority. Producing table chickens also requires registration with the Environmental Health department. Livestock welfare, food safety and packaging legislation will also need to be complied with. In the USA, processing poultry for sale requires registration with the US Department of Agriculture (USDA).

The traditional method of plucking is by hand. The skill lies in removing the feathers without tearing the skin. There are small machine pluckers on the market. Wear an overall, head covering and if necessary, a face mask. A well ventilated barn or outbuilding with a sealed, concrete floor makes a good plucking area.

To eviscerate a bird, use a beech table that can be scrubbed with hot water. Cut off the head and make a cut in the skin along the line of the neck. Use poultry scissors to cut through the neck and remove it. Enlarge the incision in order to get the hand in, and remove the gullet and fold over the flap of skin. Turn the bird round and make a circular incision around the vent so that it can be pulled clear. Take care not to pierce the rectum. As the vent is pulled clear, the intestines follow after, and can be dropped into a bucket. Enlarge the opening enough to get your hand in and draw out the gizzard, liver, crop and lungs.

Next, make an incision just above each foot, but do not cut right through. Break the leg bone at this point by snapping it over the edge of the table, and the leg tendons will be revealed: they look like white elastic bands and are very tough. The easiest way of removing them is to have a purpose-made tendon remover screwed onto the wall: it looks like a double hook, and can be bought from equipment suppliers. Place the foot between the prongs and then pull the carcase down sharply, and the foot and tendons will be pulled away. Then store the bird in a cool room, breast-side up, until it is required for cooking or for sale.

Eggs

Eggs should be collected regularly and stored in cool conditions until used or sold. If nest boxes are clean and well managed, there should be few, if any, dirty eggs; if there are, they should be cleaned and cooked for home use without delay. Only clean, unwashed eggs should be sold. These are the equivalent of Grade A eggs, the top quality bracket, in which free-range eggs are included. Surplus eggs can be sold to friends and neighbours and to callers, as long as they are not graded into sizes or sold to a shop or hotel. In these cases, registration is required, as well as

Egg Sizes			
European Union		*USA*	
Very large	70g and over	Jumbo	2.5oz
Large	63g up to 73g	Extra large	2.25oz
Medium	53g up to 63g	Large	2oz
Small	Under 53g	Medium	1.75oz
		Small	1.5oz
		Peewee	1.25oz

regular testing for internal and external quality (*see* Reference section).

Like all birds, chickens moult each summer, with some old feathers being dropped and replaced by new ones. They will often cease laying at this time, and commercial producers will normally have a younger flock to take over egg production and make up the deficit. Ensure that the moulting hens have adequate protein in their diet (for feathers are mainly composed of protein), and allow them to enjoy their holiday from production while they are re-feathering. They deserve it!

The egg-laying system is also influenced by the amount of available daylight. Once daylight hours decline, hens may cease laying until the spring, though this is more likely with the older breeds. Hybrids, and particularly those that were hatched in the spring, will tend to lay through the winter – and if the hens are provided with just enough artificial light to make up the deficit of natural light, they will almost certainly be stimulated into supplying eggs throughout the winter. There are quite sophisticated systems now available for the small poultry keeper; these have an electronic sensor that automatically 'reads' how much natural light is available, so no manual adjustments are necessary: the light, powered by a 12v battery, turns on and switches off automatically, so there is no need for an adjustable timer in the circuit.

Coping with Problems

A great deal can be done to avoid problems. Bought-in birds should ideally have been vaccinated against Marek's disease, Newcastle disease and infectious bronchitis, unless they are table chicks destined for organic production. If bought from a commercial rearer, they will also be from a salmonella-tested flock. Newcastle disease (fowl pest) and avian influenza are notifiable diseases, meaning that their presence must be notified to the authorities.

Regular cleaning of houses, the availability of fresh pasture, good quality food and clean water will go a long way to keeping birds healthy. However, there are always problems, although small flocks in non-intensive situations are less likely to encounter them than larger ones.

There is a range of bacterial and viral diseases that can affect chickens, but veterinary advice is usually necessary to diagnose them. Bacterial infections can be treated with antibiotics but any eggs should be discarded during the course of treatment, and for a withdrawal period of several days afterwards. Viral infections cannot be treated, which is why vaccination against the more serious ones, referred to above, is important.

It is worth remembering that cryptosporidium, salmonella and campylobacter are all organisms that can affect humans, so it is important to wash your hands after handling poultry or their housing. Newcastle disease can manifest as a mild form of influenza in humans. Psittacosis, usually referred to as parrot disease, can also be transmitted to humans from chickens. It is a serious condition, but is extremely rare, particularly as free-range chickens are not confined to indoor bird rooms where dust accumulates.

It is not necessary to trim the beaks of chickens that are kept in small flocks and are under good management.

External Parasites

Lice and mites can be a problem, especially in summer. Severe infestations can cause anaemia and even death. The red mite is a particular nuisance because it hides in cracks inside the house during the day, coming out at night to prey on the birds when they are roosting. Suspect its presence if hens suddenly show a reluctance to go in at night, where previously they have perched quite readily. Several proprietary products are available, including natural biocides that can be used in an organic unit. Some of these are based on pyrethrum from the chrysanthemum plant.

A particular nuisance is the scaly leg mite, which pushes up the scales of the legs with the encrustations that it produces. The legs must be soaked in warm, soapy water to dislodge the crusts; do not pull them off, otherwise the legs will bleed. Once the legs are clean and dry, they can be treated with an anti-mite preparation.

Internal Worms

Free-ranging birds can pick up worms, particularly if pasture is not adequately 'rested' to break the life cycle of pests. Flubenvet is a wormer that can be added to the feed and is suitable for all poultry. It needs to be fed over a period of several days. A convenient time to administer it is when the birds are coming to the end of their moult in summer, and before they start laying again.

Pests and Predators

Vermin are defined as animals and birds that are troublesome to man and domestic animals, but it is important to remember that a troublesome creature is not always vermin. In Britain, for example, all birds of prey and owls are protected by law, although crows, rooks, magpies and jackdaws can be shot during the day or caught in live catch traps; however, they may not be snared, spring-trapped or poisoned.

The Fox

The fox is undoubtedly predator number one when it comes to free-range poultry. The only security is to fence it out with high fences or electric fencing, as detailed earlier. There are also large cages that the animal enters and cannot get out of again – but it then poses the question of what to do with it. The most humane method of despatching it is with a 12-bore shotgun, as long as the owner holds a valid firearms certificate.

An ordinary flashing light is a deterrent as long as its position is moved fairly frequently. However, once the fox gets used to it, it is no longer frightened away. Such units are available for use with electric fencing, or they can be incorporated into a normal fence.

Chemical deterrents are available that act as a repellent when used along boundaries; Renardine, for example, can be mixed with sand and strewn along the perimeter. The treatment needs to be renewed weekly, or after heavy rain. A family dog may keep a fox away, while lion or tiger dung from the nearest zoo, placed along boundaries, is said to be effective. Llama keepers are united in the view that llamas will drive off foxes.

Rats

The brown rat, *Rattus norvegicus*, is a major pest because they are so numerous. They eat and contaminate stored feeds, kill young birds and carry disease, including salmonella bacteria, trichinosis nematodes, toxoplasmosis protozoans, and a range of viruses. From man's point of view, the most serious threat is Weil's disease. (If there is any evidence that rats have been around, it is important to wash the hands after touching any surfaces where their saliva, urine or droppings may have been deposited. Treat any cuts and grazes with antiseptic, and keep them covered.)

If you see one rat, you can be certain that there is already a colony nearby. Their average foraging distance is usually no more than 50m (165ft) from the nest. Having once established routes from the nest to the food source, they will travel along the same 'runs' each time they visit. This tendency makes it relatively easy to control them. Try to identify possible hiding places. Are there any

Rats are intelligent and adaptable creatures, and are a health hazard to poultry and people. Photo: Sorex

Common Problems

Problem	Solution
Floor-laid eggs	Ensure that nest boxes are easily accessible. Remove straw from other areas.
Dirty eggs	Provide clean litter in nest boxes. Keep mud outside the house to a minimum.
Egg eating	Use 'rollaway' nest boxes or hang plastic strips across entrance to nest box. Try and isolate the culprit. Place blown egg with mustard in the nest box.
Feather and vent pecking or cannibalism	Check birds and houses for external parasites. Ensure feed has enough protein. Avoid mixing new and old birds. Consider getting a more docile strain. Spray feathers with a proprietary product to impart a nasty taste for attackers.
Unwanted broodiness	Remove hen and place her in cool conditions.
Impacted crop	Administer a little cooking oil and knead the crop to get rid of the blockage. Avoid stringy greens and grasses.
Sour crop	Crop feels watery to the touch and hen is not eating. Give her some fizzy mineral water to reduce the acidity. When she starts eating, mix some live yoghurt with her feed.
Fertile egg	Don't let the cockerel run with the hens.
Egg with green yolk	Remove shepherd's purse plants and acorns from pasture area.
Lameness	Check feet for wounds. Bumblefoot is a hard compacted area which has healed over but still has pus inside. It may need to be lanced. Disinfect the wound. If other birds are affected, consult the vet.
Egg-bound hen	Egg is stuck. Put Vaseline around the rim of vent. Hold hen above steam to relax vent area, but beware of scalding. If egg is still stuck but can be seen, puncture it from the outside and remove it all. This is a 'last resort' solution where everything else has failed.
Soft-shelled egg	Ignore unless it is a regular event, or several birds are affected. If so, contact the vet. Ensure that there is enough calcium in diet.
Feather loss	Natural, seasonal moult. If excessive or prolonged, check for external parasites or possible sources of stress. Ensure that food has enough protein.
Prolapse	Vent area hanging out. More common in heavy layers. Push back and treat with antibiotic from vet. Severe cases should be culled.

identifiable 'runs' indicated by flattened areas of vegetation or soil running along the edges of buildings or other structures? Cut down tall weeds growing against buildings, and clear areas where piles of timber or flagstones may have been left for some time; they often conceal rat runs.

Inside outbuildings, paint a white strip along the floor perimeter to show up the presence of droppings more easily: the spindle-shaped droppings are quite large, up to 2cm long. Look out for evidence of gnawing. All feeds should be stored in a rat-proof building or in a strong container with a tightly fitting lid; dustbins are a good option for low-cost and effective storage.

There are Fenn traps for killing rats, but placing them is a skilled job, ensuring that they are in the runs and not accessible to anything else. Poisoning is the most common option, but again the bait needs to be distributed carefully so that it is accessible to the rats, but not birds, domestic pets and, of course, children! The main rodenticide used is an anti-co-agulant. The baits are based on cereals and are available loose, in sachets or in blocks. The general rule, as far as dosage is concerned, is that if all the bait is taken on the first setting, double the amount is placed the second time, and so on; this ensures that *all* the rats are being dealt with.

Having cats and a dog may help to deter rats from a site, but once established nearby, their presence will not control them to any measurable degree. The obvious exception is the Jack Russell, which was bred as a ratter. Even so, it is usually necessary to dislodge the lair in order to drive out the rats before the Jack Russell can get to work.

Ferrets can be very effective in going into a lair to kill and drive out rats, but they need to be trained for this kind of work; many, these days, are kept purely as pets. It has been said that the best combination for dealing with a lair is four ferrets, two Jack Russells and six men (or women), with the ferrets providing the inner circle, the dogs the next one, and the humans on the outer circle to deal with the last escapees. Long trousers tucked into boots are essential, because it is not unknown for a frightened rat to run up a trouser leg.

Mice

House mice do not carry as many diseases as rats, but they can be a nuisance. If there is an infestation in a dwelling, their gnawing habit can be a fire hazard if they attack cables. They will also get into and despoil feed stores. As for rats, mice may be poisoned, or they may be trapped, either with the snap-bar type that kills the animal on entry, or the 'humane' one that catches it alive.

Cats, particularly those with Siamese in their family tree, make good mousers, and females are often the best choice because they have less of a tendency to wander than males.

Wild Rabbits, Hares and Deer

These animals do not present a particular problem for poultry, but they can be a nuisance where crops are concerned. The best approach is to fence them out. Standard rabbit fence has the netting well buried and turned outwards at the bottom. Deer fencing, however, is an expensive operation. Trees may require tree-guards to protect them. Electric fencing is also effective.

Mink, Weasels and Stoats

Again, the best approach is to fence them out of where poultry are kept, but all of these can get in through small apertures, such as the gaps in a netting fence. A proprietary repellent is sometimes effective, also soaking rags in diesel oil and hanging them on the boundary has been known to work.

Hawks

Overhead predators such as hawks can prey on small fowl, as well as frighten large ones. There are raptor deterrents available in the form of a ball mirror glass, and ordinary mirrors positioned to face upwards are also effective. Birds of prey can also be deterred by the use of scarecrows, as long as these are moved fairly frequently.

Feral Cats and Larger Predators

Sometimes feral cats or other people's pet cats may be a nuisance if they come into the garden. A hen run can, if necessary, have garden netting put over it to keep them out. The most effective deterrent, however, is to encourage your own cat or dog, or both, to stake a territorial claim that will be heeded. Domestic cats and dogs will usually learn not to chase poultry or other stock on their own territory.

Normal fencing may not always keep out larger predators on a field scale, but electric fencing almost certainly will.

11 Rabbits

The care is all; and the habit of taking care of things is, of itself, a most valuable possession.
(William Cobbett, 1822)

The rabbit, *Oryctolagus cuniculus*, was known to the early Phoenicians, but there is no record of their having been introduced to Britain before the Norman invasion. The idea of keeping them in hutches is a relatively recent idea. From medieval times to the eighteenth century, the normal practice was to have a warren. This involved having a large enclosure where burrowing was encouraged, often by the drilling of artificial burrows. Here, the rabbits lived a semi-wild existence and could be trapped as required.

Rabbits are quiet creatures, gentle and amenable in their disposition. They share no diseases with humans, and are the most popular pets after cats and dogs. They fit in well with other smallholding activities, for rabbit breeding is an activity that can be carried out on any scale. There are no restrictions on keeping rabbits, other than the normal requirements for animal welfare.

Breeds and Types

It was not until the end of the eighteenth century that specific breeds were recorded by name in Britain: these were the Spotted English, the Lop and the Angora. In the following century, exhibiting began and many breeds such as the Himalayan, the Beveren and the Dutch appeared. Since then, many varieties have been introduced. There can be a considerable variation in type of fur as well as size, from the tiny Netherlands Dwarf to the enormous Flemish (British) Giant.

Fancy: Generally speaking, these are animals that have been bred for show features such as size, type of ears, and colour or texture of coat, rather than for utility purposes. Examples include the Lops, Netherlands Dwarf, Flemish Giant, Belgian Hare, Angora, Harlequin and Magpie.

Tricoloured Dutch rabbit. Note the nipple drinker attached to the hutch door.

Normal Fur: Normal fur breeds have an undercoat that has guard hairs projecting above it; examples are the Beveren, Californian, New Zealand White, Sable, Chinchilla and Siberian.

Rex: Rex breeds have guard hairs that are either the same length or shorter than the undercoat, so they are not visible; the overall effect is a short and plush coat. Originating in France, they were originally developed for the commercial fur trade. They are now available in a range of groups such as Self, Shaded and Tans: this refers to colour and patterns.

Satins: Originally bred in the USA, the Satin breeds have hair that is quite different from that of other groups, a mutation of the hair structure giving the coat its characteristic satin-like sheen and texture. Examples include the Argente Champagne and Havana.

Show and Pet Rabbits

There is an overlap between pet owners and show people, but the biggest sector in rabbit keeping is undoubtedly the pet world. The British Rabbit Council produces a Standards Book, giving details of the categories and breeds that are recognized, while in the USA, the relevant organization is the American Rabbit Breeders' Association. In Britain, pure-bred show rabbits are normally identified by having a leg ring with a unique number that is provided when the rabbits are registered. In the USA, identification is by ear tagging or tattooing.

Breeding rabbits on a small scale to supply local pet owners and pet shops is a possibility that is worth investigating. There is also a growing number of people with 'house rabbits' – pet rabbits that live in the house and garden, just like the family dog or cat. It is possible to house-train a rabbit to use a litter tray, just like a cat. Rabbits, guinea pigs, cats and dogs will all cohabit harmoniously in a domestic setting, as long as they are introduced to each other when they are young, and grow up together.

It is quite common to keep a buck in an outside hutch and run, while females are house-trained for inside occupation, and then take it in turns to visit the buck when litters are required. They would also

Dwarf rabbits are particularly popular as pets.

have their own hutch and run outside.

Whichever breeds are chosen, it is a good idea to join the appropriate societies. Rabbit shows are held at many of the country shows, and this is also a good opportunity to see the animals and talk to breeders. Breeders who are supplying the pet shops and garden centres insist that quality pays, because customers will pay more for a rabbit that is well handled and trained; of course it is also essential to ensure that animals are healthy and in good condition.

No rabbits should be sold before they are eight weeks old. It is also in everyone's interests to ensure that the rabbits are good examples of their breed, and that the breeding males and females are from different lines, so that congenital defects from too much in-breeding are avoided. Start with the best breeding stock available, and if they are show breeds, register them with the national association.

Table Rabbits

Although all rabbits, including the wild one, will provide meat, there are some breeds that have been developed with greater meat-to-bone ratio, which will grow more quickly, and which are therefore more suitable for the table.

New Zealand White: The main meat breed is the New Zealand White. Originally developed in the USA, it is now widely available. It is an albino with pink eyes and puts on weight rapidly. Commercially, it has been developed to produce rabbits with a live weight of 2–2.2kg (4.5–5lb) in nine to ten weeks.

In the USA, the Florida White is a smaller rabbit based on the New Zealand White, while in the UK, the Hybrid White is a selected strain that has been developed for the table trade.

Californian: Also developed in the USA, the Californian weighs between 3.6–4.9kg (8–11lb) as an adult, but is slower growing than the New Zealand White. The skin is preferred by many however, because it is thicker and more substantial. As well as having a commercial value, the Californian is popular as a show breed. It is white in colour, with ear markings (points) of either black or chocolate brown.

Fibre Rabbits

Angora rabbits are kept for wool production, with the longest-haired strains being the most valuable. They are frequently kept in cages so that litter does not stick to, or stain the fur. The coat needs regular brushing to avoid matting, a procedure that can

Californian rabbits are raised for meat, but are also popular as pets or for shows.

Angora rabbit, a breed kept for its fine wool.

begin as early as five weeks of age and must never be neglected.

Commercial or Continental strains of Angora are bigger than show Angoras. The English Angora weighs about 2.7kg (6lb), with an annual yield of wool of around 350g (12oz). The German Angora is the largest breed, weighing up to 5kg (11lb) and producing up to 1.2kg (2.6lb) of wool. French Angoras can weigh up to 4.25kg (9.4lb), with an annual yield of 1.2kg (2.6lb). Most Angoras kept for their wool are white, but there are also coloured varieties; these include Smoke, Golden Fawn, Sooty Fawn, Cream, Blue Cream, Brown Grey (Agouti), Blue-Grey, Chinchilla, Sable, Chocolate and Cinnamon. In the USA, there are Giant Angoras, as well as Satin Angoras.

Commercial mills are normally interested in white wool only, and in large quantities, so the small producer is best advised to cater for the home crafts sector. The fine wool can be spun to produce yarn; finished garments may be knitted, woven or crocheted, and felt provided for sale.

Angora does are mated at around six months, and are normally allowed two litters a year. Commercially, they are clipped at intervals of around fourteen weeks. They are first clipped along the back, then underneath and along the sides, taking care not to cut the animal. In France, the practice is to pluck the wool when the fleece is well grown and the animal is moulting naturally; a small comb makes the process easier, and it certainly does not hurt the rabbit.

The wool is separated and graded according to length and quality, though different countries have their own grading systems. In Britain, one system in use is as shown in the box.

Housing

The choice of housing is between hutches and cages. The Royal Society for the Prevention of Cruelty to Animals (RSPCA) recommends the following minimum dimensions for a hutch to house two small or medium-sized rabbits: 150 × 60 × 60cm (60 × 24 × 24in). For two large rabbits, the minimum dimensions are 180 × 90 × 90cm (72 × 36 × 36in). Hutches are available as indoor or outdoor versions, as well as being single or multi-tiered. They should be well clear of the ground to avoid damp. Outside hutches need a weatherproof roof that is angled away from the front, to shed water, and which overhangs the hutch to provide extra protection. They should be sited in a protected area, facing away from the prevailing winds.

Hutches are normally divided into two sections, with a sleeping area leading off from the larger living area; each hutch will need a feeder and a drinker. Bottle drinkers fitted with a tube attachment (nipple drinkers) are convenient, and the rabbits soon learn to drink from the tube; this is more

Grades of Angora Wool		
Grade 1	A:	Clean, non-matted. Minimum 60mm long.
	B:	Clean, non-matted. Over 30mm long.
Grade 2:		Wool from young rabbits in first shearing or short adult wool, under 30mm.
Grade 3:		Clean, matted wool.
Grade 4:		Dirty, matted wool.

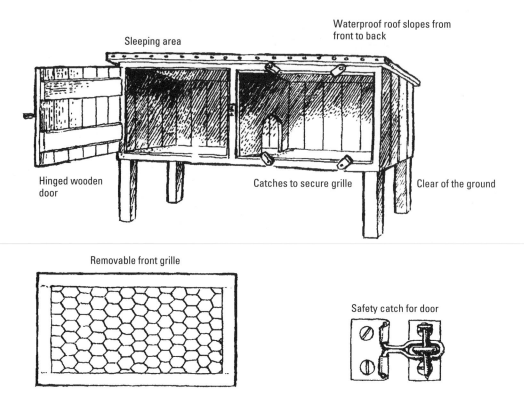

A typical rabbit hutch.

hygienic than some of the traditional drinkers where droppings may contaminate the water. A hayrack is useful for feeding hay and greens.

Chopped straw can be provided in the hutch. Bags are normally available from livestock supply retailers and will be clean and free of mites, as well as toxic residues. Litter absorbs droppings and urine and helps to keep the hutch clean, but it needs to be replaced regularly. A dustpan and brush is ideal for the periodic sweep-out into a bucket. Rabbit droppings are a valuable addition to the compost heap or wormery; areas of compacted droppings may be removed with a triangular paint scraper. Some hutches have a droppings board that slides out for easy maintenance.

Commercial table rabbits are normally housed in cages in a well ventilated barn. Droppings fall through onto straw below, from where they are periodically removed for composting. One unit I visited had earthworm beds constructed underneath, so that a worm-farming enterprise was run in conjunction with the rabbits. For a commercial venture, automatically refilled drinkers are common, with a header tank and ball valve to control the supply to individual nipple drinkers. There are also cages available for domestic

Hutch cleaning equipment.

rabbits; these have plastic trays underneath that are easy to take out and clean.

During periods of fine weather, rabbits can be put in temporary grazing arks so that they have the benefit of grazing on grass. The provision of a complete enclosure is common for pet rabbits. Part of the ark can be covered to provide night or shower protection. Only females or a mother with young can be housed together in this way. The buck needs to have his own quarters, otherwise indiscriminate and

A tiered rabbit hutch with a run provides much more exercise area than a normal hutch.

A colony of rabbits with individual hutches and a shared exercise area indoors.

possibly harmful breeding will take place, as well as fighting. This system worked well for us, because we dislike cages. Each rabbit had a hutch in an outhouse that had electric light. They were there in periods of wet or severe weather, but had the benefit of outdoor grazing in fine weather.

Feeding

The rabbit is a gnawing herbivore with a digestive system that is adapted for fibrous food. The front teeth grow continuously, so it is vital for it to be constantly gnawing on something, to keep them in trim. Indeed, unless a rabbit is provided with something to gnaw on, it may decide to nibble its house. A small, bark-covered log is a good idea (though avoid those from evergreens); carrots are also popular.

A good quality, additive-free, proprietary feed provides most of the required nutrients. This is available either as pellets, or as a dry mix of various cereals. Additive-free feeds are available for those who wish to avoid in-feed medications. A mix suitable for domestic rabbits that have access to other foods is usually around 13 per cent protein. Pet rabbits, for example, can be given additional small amounts of crisped potato peelings, bread, garden greens and roots such as carrots. A balance of green food is better than a surfeit of one type in case it leads to digestive problems. Suitable green foods include cabbage, kale, chicory, clover, dandelions, yarrow, parsley and shepherd's purse. The provision of hay at all times provides a good balance. Pellets for commercial rabbits are around 17 per cent protein; adult rabbits, breeding stock or commercial table rabbits will need an average of 115g (4oz) pellets a day. Younger rabbits eat proportionally less. Feeding half in the morning and the rest in the afternoon is the usual practice, although some do feed the whole ration in one go, perfectly successfully.

Rabbits have a habit known as pseudo-rumination, in which they take small, soft pellets of partly digested food from the anus, which are then eaten and digested. The final waste matter is in the form of hard pellets that are then voided. This pattern of behaviour is quite normal.

Breeding

There is no period of 'heat', as there is with other animals; the doe simply ovulates in response to the buck. When breeding is required, she should be put in the buck's quarters, and never the other way around or she will attack him. (The correct way to lift a rabbit is to hold it by the scruff of the neck and to support its bottom; the ears are delicate and should never be used to lift it.) Usually, mating will take place within a matter of minutes, but sometimes it happens that the doe runs about and will not stand for the buck. In this situation the best solution is to conduct an 'assisted mating'. This involves holding the doe in a stationary position, with one hand underneath her abdomen so that her hocks are slightly raised, while the other hand holds the scruff of her neck so that she cannot escape. In this way, mating will occur quickly. Both buck and doe should be at least six months old, because this allows time for their characteristics to be assessed. They should obviously be in perfect health, free from physical deformities, and unrelated.

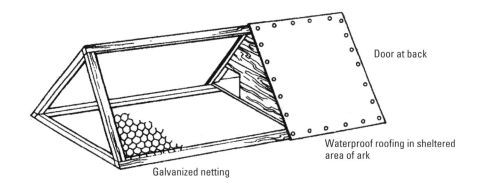

Door at back

Waterproof roofing in sheltered area of ark

Galvanized netting

Grazing ark.

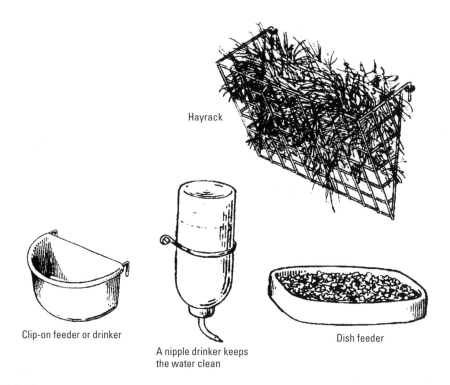

Hayrack

Clip-on feeder or drinker

A nipple drinker keeps the water clean

Dish feeder

Feeding and drinking equipment.

Depending upon the number of does, a breeder will need to work out approximately how many litters per year are required, so that the total number of rabbits in a year can be estimated. Commercially, each table doe produces an average of six litters a year, and the breeder will aim for a minimum of nine baby rabbits per litter. Fibre rabbits have two or three litters a year. Small breeders will normally opt for less frequent breeding.

The keeping of records is important, and each doe should have her own record card. This will give details of her breed and number, the date and number of the buck she mated with, followed by the date of 'kindling', or giving birth, with the number of live and dead births.

The doe's pregnancy will last thirty-one days. For the first twenty-three days she will require a maintenance allowance of 115g (4oz) of feed pellets

An ideal environment for rabbits. The hutch provides shelter and security, while the hollow log is an excellent play area. The immediate area is fenced off against predators.

Two-week-old New Zealand white rabbits.

Alert disposition

Bright, clear eyes

Nose free of discharge

Ears free of orange deposits that indicate presence of mites

Sleek, glossy coat

Well-fleshed without being over-fat

Scut area free from diarrhoea

Feet free of wounds without over-long claws

What to look for in a healthy rabbit.

or cereal mix a day. On the twenty-fourth day, her feed allowance should be doubled to 225g (8oz) to take into account the forthcoming lactation. She should always have clean, fresh hay to nibble. Some people also feed raspberry leaves at this time, for the beneficial effect on the uterine muscles during labour.

Three weeks after mating, she can be given a nest box with some chopped straw or hay (though if she is in a hutch, her sleeping compartment will suffice); she will make a nest in this. When she is close to giving birth, she will pull out some of her own fur to line it. At this stage it is useful to install a litter board, a strip of wood about 7cm (3in) wide, fitted along the front of the hutch, to prevent the babies or bedding falling out when the door is opened.

The doe is best left to her own devices during the actual birth, or 'kindling', and left undisturbed until all the young have arrived, been cleaned by her, and covered up in the nest. Often the first indication that anything has happened is when you suddenly notice a movement in the nest. Closer examination will reveal several pink and hairless babies with their eyes closed. Make a careful examination to check that the doe is all right, and to count and record the number of young. Any stillborn or physically deformed ones should be removed and disposed of.

The mother will periodically hop into the nest to feed the little ones. While they are tiny, she will carefully cover them over each time, to keep them warm until their fur grows. As they grow, they become more adventurous, and if the weather is mild and sunny, they and the mother can go out together in a grazing ark.

The doe will continue to feed them until they are about seven weeks old, when they will be weaned. Before this, they will already have begun to gnaw solid food, and it will be necessary to provide extra rations to cater for them. Once they are weaned, the doe can be moved to her own quarters; and at around ten weeks old, the young should be separated from each other, otherwise they will fight.

Health

Few serious problems should arise with good, healthy stock, if they are properly fed and watered, and if they are housed in well ventilated conditions.

Vaccinations

There are two diseases against which breeding stock and pet or show rabbits should be vaccinated: these are myxomatosis and rabbit haemorrhagic virus disease (RHVD). Both are highly infectious and are killers. A veterinary surgeon will carry out both vaccinations on rabbits from six weeks old, and will provide vaccination certificates for them.

Mites

Ear canker mites can be troublesome: the rabbit will twitch its ears excessively and rub them with its paws if mites are present, and an examination will reveal orange-coloured secretions in the ear passages. These should be carefully removed with cotton buds dipped in warm, soapy water, and a few drops of veterinary canker liquid administered. It is a good idea to put some more drops in the ears a week later to ensure that all the mites, plus any that might subsequently have hatched, are destroyed.

Mange

Also caused by mites, this is a condition where bare, often sore, patches appear on the skin. Veterinary preparations can be administered, not only to kill the mites, but also to soothe the skin.

Fleas

Like all warm-blooded animals, the rabbit is subject to flea attack. Check their pelts regularly, and if fleas are detected, treat them with an anti-flea preparation.

Over-Long Claws

The claws will need regular clipping: since the rabbit is not burrowing, they will become long and overgrown. If the foot is squeezed slightly, the claws will protrude, and the position of the blood vessel in each one can be established by looking at it in a strong light. Clip the claw back, ensuring that the cut is well clear of the blood vessel. Ordinary manicure clippers are quite satisfactory.

Hairball

Sometimes self-grooming results in a hairball forming in the stomach; this occurs more commonly in Angoras than in other breeds. It is important to brush out the loose hair of long-haired breeds regularly. In the event of a hairball, an accepted remedy is to give pineapple juice to drink on three consecutive days, as it contains an enzyme that assists in the breakdown of the hair. Give only good quality hay to eat during this period.

Snuffles

Watering eyes and a runny nose may mean that a rabbit has a cold, usually referred to as 'snuffles'. A little onion and garlic added to the food will help, but there must also be freedom from draughts and damp, and good ventilation. Like our common cold, it usually clears up.

Pasteurellosis

Badly ventilated and unhygienic housing is the most

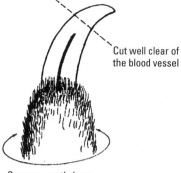

Cut well clear of the blood vessel

Squeeze gently here

Clipping a rabbit's claws.

frequent cause of this airborne infection. Outdoor rabbits rarely get it. It can be treated with antibiotics.

The Blows

This is a digestive condition in which gas collects in the caecum, and the rabbit sits in a huddled-up position. It is usually caused by giving too much green food, particularly brassicas. Feed only hay, a few pellets and water until the condition clears up.

Coccidiosis

This is a serious disease that leads to liver infection, listlessness and diarrhoea. It is caused by a protozoan that is normally picked up from damp litter, so it is vital to ensure that a build-up of such material does not occur in hutches. It can be cured with antibiotics from the vet.

Mastitis

This is an infection and inflammation of the teats, in which they become sore and hard. It is important to watch out for signs of mastitis after a doe has kindled. An antibiotic from the vet may be necessary.

12 Waterfowl

I can honestly say that I know of no livestock which can prove of more interest or give so much pleasure.
(Reginald Appleyard)

The Importance of Water

All waterfowl share a common need for water in their environment. The source of water may be a farm pond, a brook running through the area, a tank that can be easily moved and cleaned, or a small garden pond. The most important factor to bear in mind is that unless the water is kept clean and oxygenated, it is a potential hazard. Stagnant water, with a build-up of decaying matter, a lack of oxygen and a warm temperature, provides the perfect conditions for botulism organisms to flourish. Waterfowl dabbling about and straining small organisms from the water in their bills can, and do, succumb to pathogens.

If a new pond is to be made there are several options, using materials such as concrete, puddled clay, rigid plastic or a plastic pond liner. It may also be possible to rehabilitate a stagnant old farm pond. The chosen site needs to be level, and without overhanging trees that would shed their leaves into the water.

Concrete Ponds

Making a concrete pond is the most costly and time-consuming method. A hole is excavated to the required depth, ensuring that there is a ramp for ease of access into the water. If it is at least 90cm (3ft) in the middle, it is not likely to freeze right through in winter. Concreting needs to be carried out in frost-free conditions, and the shell must be rendered and left to weather for a few weeks before being filled with water. It provides a strong and durable structure, though some means of aeration is required if the water is not to become stagnant.

On a small garden scale, where only a couple of ducks are involved, it may be more appropriate to use an old-fashioned, deep sink. Sited over a soak-away hole filled with stones, it is then merely a matter of pulling out the plug to let out the dirty

The pond at Broad Leys. It catered for ducks, fish and plants, as well as pond dipping activities by the author's youngest son Gwilym.

A pond in the course of construction, but the birds have already taken up residence. Photo: H.D. Sharman

A children's paddling pool adapted for ducks by having a ramp attached to it.

water and refilling it as necessary. On a larger scale, refilling and outlet points will be required so that regular replenishment is possible. Alternatively, a pump can be used to provide aeration.

Puddling a Pond

This is the traditional method of making a pond, and it requires a large amount of natural clay. It is more appropriate for soil that is naturally heavy. A load of clay can be delivered by contractors, but it is an expensive option. The clay is placed in an excavated hole and then 'puddled' or stamped down until there is a sealed layer about 25cm (10in) thick. A layer of soil is then placed on top, raked smooth and allowed to settle.

Rigid Plastic Ponds

Pre-built, rigid structures are widely available in various sizes. It is a matter of digging a hole and then lining it with sand before inserting the pond. It needs to be an exact fit otherwise the structure may crack. Plenty of sand is needed to accomplish this.

There are purpose-made duck pools that are fairly shallow but are equipped with taps for draining and refilling. The latter are normally sold by poultry equipment suppliers. On a small scale, a child's rigid plastic paddling pool works well: it is light and easy to move, and can be easily emptied and refilled. Such a structure will need to have a sloping ramp provided for access.

Pond Liners

The use of pond liners simplifies the whole process of construction. Butyl liners are the toughest available. A hosepipe or rope can be laid down to mark out the outline of the proposed pond, and then

a hole excavated to the required shape and depth. Sharp stones need to be removed in case they puncture the liner. The hole is then lined with underlay or sand, and the pond liner placed in the hole, the edges anchored with stones. Gradually fill with water, folding and pleating the liner into place. Finally, trim the surplus liner, allowing sufficient overlap, and place paving slabs around the edges.

Pumps

Pumps are of two kinds: low-voltage submersible, and mains surface. The former is where the pump is put in the pond, set on a platform above the silt layer, and connected by low-voltage, heavy-duty cable to the mains via a transformer; this steps down the mains current to 12v or 24v. The lead from the pump is connected to the low-voltage cable from the mains with a waterproof connector on land. It is a good idea to bury it and place it under a flagstone, for example, so that it is protected but accessible. The cable can be run through plastic conduit pipe for added protection. A low-voltage, submersible pump is a cheap option and will effectively power a fountain. It is easily installed and does not require an electrician.

A mains surface pump is usually housed in a small pump chamber near the pond, but clear of it. It is a more expensive option and needs a qualified electrician to install it.

A recent innovation has been solar-powered pumps. These are of two types: the first has the solar panel placed on the ground near to the pond, with the appropriate wiring going to the pump. The second is a complete unit that floats in the water, coming into action when the sun shines on it.

If green blanket weed is a problem, put some barley straw in a netting bag weighted down with stones, and immerse it in the pond. If this is done from late winter onwards, the toxins released by the straw decomposition have an inhibiting effect on blanket weed, without affecting either fish or ducks – in fact, the straw provides a home for invertebrates that provide them with a tasty food source.

Ducks

Four ducks on a pond, a grass bank beyond.
(William Allingham, 1870)

All breeds of domestic duck (with the exception of the Muscovy) are descended from strains of the wild mallard, *Anas platyrhynchos*. The Muscovy, *Cairina moschata*, which is more like a goose in some of its characteristics, is generally acknowledged to have evolved in South America as a distinct species.

Ducks are beautifully designed for the water. Their general shape, with the centre of gravity in the keel, as in a boat, enables them to maintain balance, while their paddle feet send them skimming through the water. They are far hardier than chickens, and have fewer of the diseases associated with other poultry. The body is kept warm by a thick inner layer of light down feathers, while the outer feathers are waterproofed by oil from the preen gland, situated close to the tail. Despite these protections, ducks actively dislike winds, so adequate shelter is essential. They must also be protected against foxes,

although having a pond does provide a safe haven. Many people have ducks to provide winter eggs when the chickens may have stopped laying.

Breeds

Pure breeds of domestic duck have their own breed standard as a means to show how well a particular bird measures up to the ideal. For commercial use, some breeds have been crossed or hybridized for utility purposes. Whatever breed is chosen, the birds should be in first-class condition, bright-eyed, active, alert and curious. Posture, stance and balance should be good, while the eyes, nostrils and vent need to be clear of discharge.

Aylesbury: This is the traditional, all-white English table duck, originating from the town of the same name. It is fairly slow growing, and has been crossed with the Pekin in order to improve its performance. Selected strains of the two breeds were used to produce the commercial white table duck. The eggs may be white, tinted or greenish, depending on the strain.

Black East Indian (East Indie): An American breed, this is a bantam thought to have been a mutation from the mallard. It has beautiful black plumage with an iridescent greenish sheen. The eggs are dark to light grey or blue. The first eggs tend to be darker, and they gradually become lighter as the season progresses.

All breeds of domestic duck, such as the Aylesbury and Pekin, are descended from the wild mallard.

Neat, rounded head

Nostrils on bill clean
and free of discharge

Wings held against the body, with no drooping

Good carriage (depending on the
breed, this will be upright or more
horizontal)

Healthy preen gland at base of tail for
waterproofing feathers

No swelling in legs or joints. No limping

What to look for in a healthy duck.

Blue Swedish: A heavy breed, the Blue Swedish has dark blue plumage with a greenish sheen and attractive lacing on the feathers. Breeding produces some colour variation, with ducklings hatching in the ratio of 50 per cent blue, 25 per cent black and 25 per cent silvery white. The eggs are grey, blue or tinted.

Call Ducks: Originally called decoy ducks, these small birds were originally used to entice wild ducks to land on a particular stretch of water so they could be shot by hunters. These days they are popular as a show breed, with a range of colour varieties available, including white, apricot, blue-fawn, bibbed, dark silver, silver, pied and mallard. They are extremely vocal and make good 'watchdogs'. The eggs are green, blue or tinted.

Campbell: Bred in Gloucestershire from the fawn and white Indian Runner, mallard and Rouen, the brown-feathered Khaki Campbell has the best egg production record of the duck breeds. Commercial strains are more productive than show ones. Campbells are also found in dark brown or white. Good strains should produce white eggs, and if they are greenish, you should suspect some wild mallard influence in the genetic make-up.

Cayuga: Originating in New York State, the Cayuga is a large, black-feathered breed with a lustrous green sheen to its feathers. One of the problems associated with breeding them is the tendency for some to have white feathers. Selective breeding is the only solution. It is a quiet breed, producing grey or blue eggs.

Crested: These all-white ducks have a characteristic crest or top-knot of feathers on the head. There is also a bantam variety. A proportion of the young are born without a crest, and hatchability is generally low. There is a lethal gene associated with the breed, in that only those with one crest gene or two crestless (normal) ones will survive; an embryo with two crest genes will die in the shell. The eggs may be green, blue, white or tinted.

The Bali breed from the East Indies is also crested, although a certain proportion of those hatched will be without a crest. Varieties include White and Khaki and resemble Indian Runners in build, although they are heavier. The eggs are white, blue, green or tinted.

Indian Runner: The Indian Runner is well named, for it runs everywhere; it has the most upright stance of any breed. The 'White Runner' was the main egg producer until ousted by the Khaki Campbell at the turn of the century. It is available in a range of colours, including white, fawn, black, chocolate-brown, blue, and fawn and white. The eggs are slightly larger than those of the Khaki Campbell and may be white, tinted, green or blue.

Magpie: Bred in Wales, the Magpie was traditionally raised as a dual-purpose egg-and-table breed. It has attractive black and white plumage. There are also blue and white, and dun and white varieties. The eggs are tinted or blue. Good utility strains can be difficult to find, but they can undoubtedly be selectively bred for increased egg production by dedicated breeders.

Some of the author's Call Ducks in a sheltered area of the pond.

Khaki Campbell ducklings. This breed is the most prolific egg layer.

Indian Runner ducks in an enclosure through which the stream runs.

The Muscovy is descended from the wild Musk duck of South America. This one has hatched her ducklings in a disused rabbit hutch.

Magpie ducks, a breed developed in Wales.

Muscovy: Originating in South America, the Muscovy is a heavy bird that is more goose-like in its habits than a duck. Ours always spent most of their time on the pasture with the geese. They also like to perch, and in this respect it is a good idea to give them some straw bales to roost on, in a barn or other shelter. They are black and white, although there is a range of colour varieties, including chocolate, blue, silver, buff and pied. Their face is covered with rough, red skin referred to as 'caruncles', while the head feathers can be raised into a slight ridge-like crest. Commercial strains are used to produce the Barbary duck that is sold in supermarkets in Europe. The eggs are large and white, or tinted. The Muscovy has a well deserved reputation as a good mother, but the drake can sometimes be aggressive to other breeds.

Orpington (Buff Orpington): Bred in Kent from the Indian Runner, Rouen and Aylesbury, the Orpington is a dual-purpose duck with attractive buff-coloured plumage. In the USA it is called the American Buff Duck. There are other varieties, including black, blue, silver and chocolate: unlike the Buff, these have white bibs on the chest. The eggs are white or tinted.

Pekin: The Pekin was introduced to Britain from China in about 1874. It has an upright stance, and white or cream plumage. With its fast growth rate it became the main table breed in the USA, Australia and elsewhere, and was also used to improve the productivity of the Aylesbury and other commercial table strains. The eggs are white or tinted.

Rouen: The Rouen originated in France and has colours similar to the wild mallard. Although it achieves a massive size, it is slow growing, a factor that eventually led to its decline as a table breed. It is now regarded as a show breed. The eggs vary in colour, depending on the strain, to include blue-, green-, and white-tinted. The Rouen Clair is closely related, but is more upright.

Saxony: The Saxony is a German breed, bred in the 1930s as a dual-purpose egg and table breed. The plumage is an attractive apricot with buff and blue-grey. The eggs are white, and it can be quite productive.

Silver Appleyard: This was bred as a dual-purpose bird. The plumage is silvery white, fawn and blue, with a mallard head in the male. There is a bantam version. The eggs are white or tinted.

Welsh Harlequin: Bred in Wales, the plumage is similar to that of the Silver Appleyard, but it is lighter in weight. The eggs are white or tinted.

Housing

The position of a duck house is crucial if muddy conditions are to be avoided. The ideal site is on a slight slope on sandy ground leading down to a pond or stream. The so-called Dutch system imitates a natural stream. It utilizes a series of continuous troughs through which water is pumped, like a stream. A smaller version of this system can be seen at many waterfowl and farm parks today, where individual houses have runs leading down to a stream, which is itself fenced off in sections. Not

many people have a stream running through their property, but the basic principle of having a house on the highest point of a slope is a good one. If necessary, it can be placed on a concrete base that is easily hosed down. If ducks have to be kept in a relatively small area, it is often better to have the whole area concreted, with a pond set in at the lowest point. If this has a draining facility, the water can be replaced at frequent intervals.

Ducks can be allowed to roam a garden area from such a run as conditions permit. They will clear an area of slugs, and do not generally cause as much damage as chickens, although young seedlings will need to be protected. Many people let their ducks range in the vegetable garden in winter, keeping grassed areas available for the summer.

Unless ducks are confined to a house until they have laid their eggs, it is a matter of 'hunt the thimble' in trying to find the eggs. They lay early in the morning, so keeping them housed until mid-morning usually ensures that the eggs are laid inside, in an accessible place. When we first started keeping ducks we were letting our Khaki Campbells out much too early and they were laying everywhere – even on the middle of the lawn and in the pond. The sharp-eyed magpies could not believe their luck! It is often said that ducks will not use nest boxes, and indeed most duck houses are not provided with them. However, when I used an adapted poultry house for my Khakis, I reduced the height of the board in front of the nest boxes to make them more accessible, and there were always eggs in the nest boxes as well as on the floor – so obviously some of the ducks were well behaved.

A duck house can be quite simple, as long as it provides adequate cover and protection, with enough ventilation. A chicken house can be used if there is one available, and adapting it is simply a matter of removing the perch and using the door rather than the pop-hole as the exit. The latter can be used as well, of course, but ducks need a wide exit because they generally rush out in a great, packed waddle. And if the house is off the ground, there should be a ramp for them to use; their webbed feet are suited to flat surfaces, and they are not good jumpers. A house with a door that opens downwards to provide a ramp is fine.

As far as space in the house is concerned, there should be at least 0.19sqm (2sq ft) per bird, with a height of around 90cm (3ft). This allows sufficient volume of air for good ventilation. The wall area above their head level can have either a ventilation panel of wire mesh, or a few holes drilled in it. The mesh should, of course, be galvanized to prevent

A barrier across the stream keeps individual breeds to their own areas and prevents interbreeding.

rusting. In a smaller house with only a few ducks, the roof can be lower, as long as ventilation is not restricted. The requirement for space outside is the same as for chickens.

The floor may be solid, slatted or made of rigid metal mesh. The advantage of the last two is that the droppings fall through – although in my experience, neither type really keeps the floor area clean, and a solid floor is preferable in many ways. A really thick layer of wood shavings or chopped straw on it absorbs the droppings, and it is also warm. If the shavings are raked through every day and replenished as necessary, it helps to keep the eggs clean. A garden rake or a small hand-rake is ideal for this.

Where ducks are free-ranging, they will usually find their own areas of shade, in hedges or under trees and shrubs; and if the wind gets up they will waddle off to sleep in a cosy corner until the situation improves. A line of straw bales makes a good temporary windbreak in a field, while in a garden, sections of windbreak netting can be erected if necessary. Our ducks favoured the goat shelter in the field, an imposition that the goats accepted with good grace.

Feeding

The duck's mobile neck allows for sudden foraging movements when food is sighted. The bill has a rounded hook or bean on the end that helps it to catch insect larvae and other prey in the water, and along the edges of the bill are serrations that effectively strain out the water while the food is retained in the bill itself. On land their bills are used in a scooping movement, into and under the food. It is important to avoid powdery foods otherwise the nostrils can become clogged; such foods should be moistened with water until they are like crumbs in consistency. Alternatively, dry feed pellets can be given. There should always be fresh water close to the food: the ducks will shovel down some food, then waddle over to drink before returning for more food.

Compound feeds formulated for ducks are available, although they will quite happily take chickens' laying pellets. Duck feeds normally consist of wheat and maize, with soya bean meal and soya oil added in order to provide the extra protein and energy requirements. Minerals and vitamins are added to guard against deficiencies. There are also specially formulated rations for table ducks. A typical duck starter ration has 20 per cent protein, a grower ration is around 18 per cent protein, whilst a duck finisher ration is about 16 per cent protein. Organic rations are also available.

Whole mixed grain or wheat is always appreciated, but it is important to provide fine poultry grit to ensure its proper digestion. Our practice was to feed them in the same way as the hens: compound feed in the morning, and grain in the afternoon.

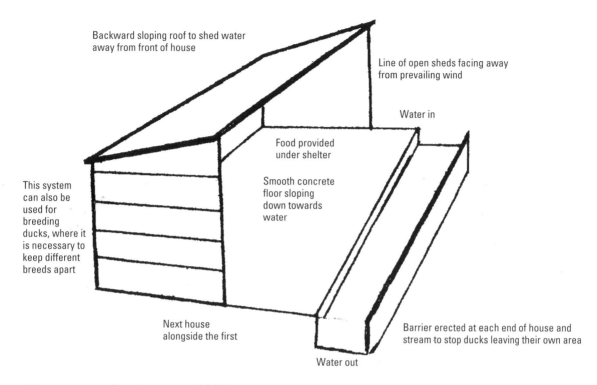

Backward sloping roof to shed water away from front of house

Line of open sheds facing away from prevailing wind

Water in

Food provided under shelter

This system can also be used for breeding ducks, where it is necessary to keep different breeds apart

Smooth concrete floor sloping down towards water

Next house alongside the first

Barrier erected at each end of house and stream to stop ducks leaving their own area

Water out

A humane system of rearing commercial ducks.

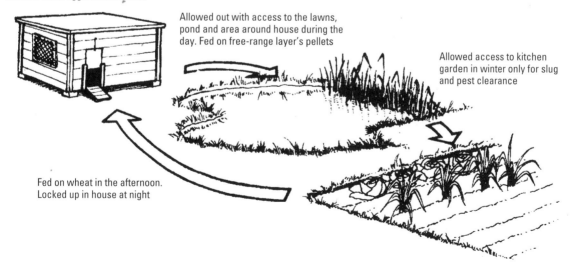

Confined in the house until mid-morning to ensure that eggs are laid inside

Allowed out with access to the lawns, pond and area around house during the day. Fed on free-range layer's pellets

Allowed access to kitchen garden in winter only for slug and pest clearance

Fed on wheat in the afternoon. Locked up in house at night

Use of land by the author's ducks.

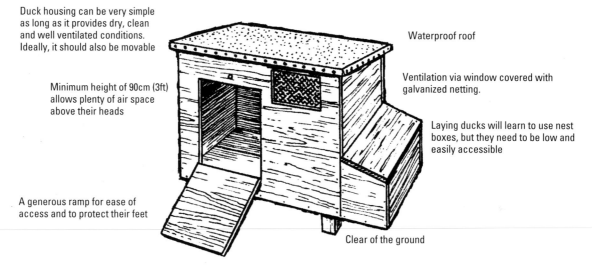

Duck housing can be very simple as long as it provides dry, clean and well ventilated conditions. Ideally, it should also be movable

Waterproof roof

Minimum height of 90cm (3ft) allows plenty of air space above their heads

Ventilation via window covered with galvanized netting.

Laying ducks will learn to use nest boxes, but they need to be low and easily accessible

A generous ramp for ease of access and to protect their feet

Clear of the ground

Small-scale housing.

Breeding

With light breeds, a mating ratio of six to seven ducks to one drake is appropriate, and for heavy breeds, four to five ducks to one drake. Before mating, they will nod vigorously to each other. They frequently mate in a pond, but it is untrue that water is required for this purpose. To avoid the possibility of low fertility and possible deformities in the young, a breeder's ration can be given to the parent birds for a few weeks before the ducks start laying. Ranging ducks will often make a secret nest and then disappear, reappearing twenty-eight days later with a trail of fluffy ducklings following behind. Broody hens can be used to incubate duck eggs, but most breeders now use incubators. The conditions are similar to those outlined for chickens, with variations as shown in the panel opposite.

Any of the brooders used for chickens are suitable for ducklings. One 250W lamp will cater for thirty to forty ducklings, depending on the breed. A duck starter ration can be given until the birds are well grown and feathered. The biggest problem with brooding ducklings is their tendency to splash water everywhere. Suspended drinkers are preferable to open ones, otherwise they will try and swim in them. Damp litter should be replaced every day. The sooner the ducklings go out during the day, the better, but they will need to be in protected conditions at night. Artificially hatched ducklings should not be allowed to go out on a pond until they are completely feathered; until then, the preen gland has not developed sufficiently to waterproof the plumage. A mother duck will normally preen her young until they are able to do it themselves, and this will give them sufficient waterproofing against wet weather and a dip in the pond.

Certain characteristics will help you distinguish the sex of ducklings: first, the females will quack while the males tend to hiss; and the males have curly tail feathers, but not the females.

Table ducks are plucked and eviscerated in the same way as chickens. If required, the down feathers can be retained and used for making duvets or pillows. On a small scale, the feathers can be sterilized by putting them in a closed (sewn) pillowcase in the tumble-dryer. The feathers are then ready for use.

Dealing with Problems

Ducks are generally healthy as long as they are fed adequately, have clean water, and clean pasture or other area in which to exercise. Bacterial infections can be dealt with by administering antibiotics. Any condition that does not clear up in a few days should be reported to a vet. Where several birds are affected in the same way, veterinary advice should be sought immediately.

Botulism

This is a lethal condition caused by a toxin produced by the bacterium *Clostridium botulinum*. It is found in dirty, stagnant water, particularly in the summer; affected birds often throw their heads backwards. The condition is usually fatal. Prevention is the best course of action, by ensuring that water supplies and ponds are kept clean and aerated.

Optimum Incubation and Hatching Conditions for Ducks			
Domestic ducks		**Muscovy**	
Day 1–24	*Day 25–28*	*Day 1–30*	*Day 31–34*
Temperature: 37.5°C (100°F)	37°C (99°F)	37.5°C (100°F)	37°C (99°F)
Humidity: 58%	75%	60%	75%

Runny Eye

Sometimes called 'white eye', this condition is caused by a bacterial infection. It may have been caused by a foreign body getting into the eye, or by the duck having insufficient depth of water in which to immerse its head. It may also be the result of having only dirty water to wash in. The eye should be cleaned with cotton wool and warm water, then treated with an antibiotic preparation from the vet. It is infectious and can be passed on to humans.

If there is also yellowish-white cheesy matter in the sac below the vent it may indicate a deficiency of vitamin A in the diet. Give small doses of cod liver oil in the food, and ensure that the ducks have access to green foods.

Lameness

A strained ligament also needs to be checked, although there is not a great deal to be done about it, but it is important to ensure that what may appear to be a strain, is not in fact a break. Ducks are prone to leg strains, which is why it is so important to provide wide, smooth access and exit points in housing or into the water.

Prolapse

Sometimes the vent will invert and hang in heavy layers outside the passage opening; however, this condition is less common than it is in chickens. Also rare is the condition where the penis fails to retract in the drake and becomes infected. An antibiotic will clear up an infection, but with any severe cases that do not improve, the birds are better culled.

Worms

Free-ranging ducks can pick up parasitic worms, not only from over-used paddocks, but from wild birds. A poultry vermifuge such as Flubenvet in the feed is effective against them. Any eggs should be discarded until a week after the course of treatment has finished.

Geese

The goose is man's comfort in peace, sleeping and waking. (Roger Ascham, 1545)

Most domestic geese are descended from the wild greylag goose; the African and Chinese breeds differ, in that their origin is linked to the wild swan goose. Geese are grazers, so the first priority is to ensure that there is enough pasture available. This will need to be good quality grassland, not over-grazed by other species, and rotated regularly so there is no build-up of disease-causing organisms or parasitic worms. Geese are selective birds, preferring new growth to lank, reedy grasses. Their bills are beautifully adapted with scissor-like serrations, just right for snipping grasses. Anyone who has ever been pinched by a gander will agree!

The grass needs to be kept relatively short, and it may be necessary to mow it regularly in order to encourage the growth of new tips. Alternatively, the geese can share a paddock with grazing animals such as sheep or cattle that will do a good job of eating down the coarser growth. If there is a choice, and a new grassland ley is being sown, one of short meadow grasses with clover is popular.

Geese need access to water for drinking and for keeping themselves clean. Domestic breeds will not require large expanses of water, but they do need sufficient to immerse their heads and splash their feathers. A small pond or accessible trough is normally sufficient for this, but the water must be clean and regularly replenished, ideally with a current of running water.

Fencing is one of those inescapable realities if geese are to be secure – even large geese are not safe from a fox. One night we were awoken by a frantic noise, to find that a fox was trying to drag a goose by the neck through a hedge. When it saw us with our dog, it dropped its prey and ran. Despite having lost most of her neck feathers and sustaining some nasty teeth marks, the goose recovered. While this was happening, the gander was lying low at the end of the orchard, disproving the claim that a gander will see off a fox.

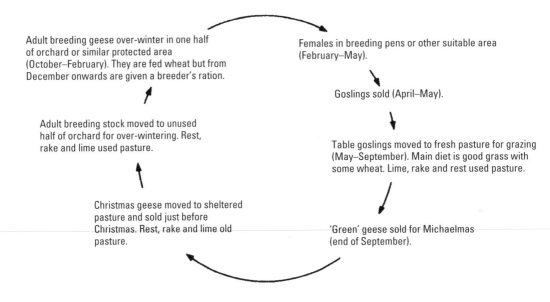

Adult breeding geese over-winter in one half of orchard or similar protected area (October–February). They are fed wheat but from December onwards are given a breeder's ration.

Females in breeding pens or other suitable area (February–May).

Goslings sold (April–May).

Adult breeding stock moved to unused half of orchard for over-wintering. Rest, rake and lime used pasture.

Table goslings moved to fresh pasture for grazing (May–September). Main diet is good grass with some wheat. Lime, rake and rest used pasture.

Christmas geese moved to sheltered pasture and sold just before Christmas. Rest, rake and lime old pasture.

'Green' geese sold for Michaelmas (end of September).

The goose year.

Breeds

For commercial purposes, hybrid Danish strains of geese such as the Legarth are frequently used because they have been developed for a higher meat-to-bone ratio. They also lay around forty to fifty eggs a season. Traditionally it was white geese of any kind that were used, with Roman and Embden usually featuring somewhere in their make-up. As far as the lighter egg breeds are concerned, the Chinese is probably the best layer.

Asiatic: Despite its name, the African goose is a heavy breed of Asian origin. This and the lighter Chinese goose are both descended from the wild swan goose; there is a characteristic knob on the head and the African has a large dewlap under the chin. The Chinese has a good egg-laying record. There are brown or grey, and white varieties of both breeds. The African also has a buff variety.

European: There are several medium-sized breeds, including the grey-back, the buff-back and the Pomeranian. Wales has produced the Brecon Buff, a breed that was once used to produce meaty birds on the Breconshire farms. The West of England breed is auto-sexing (i.e. male and female goslings have different-coloured plumage) a useful feature enabling goslings to be identified as male or female at hatching. The Shetland is a smaller version of the West of England.

The heavy breeds that provided most of the table birds include the Embden from Northern Germany and Holland. In Britain it was crossed with native white strains to increase the bodyweight before being standardized. The Toulouse from France was also developed for the table. It was crossed with native grey geese in Britain, again to increase its size. Originally a bird for the table, it is now so massive and slow-growing that it is primarily regarded as a show bird.

Light European breeds include the Roman, a white goose introduced to Britain by the Romans. The Pilgrim originated in Britain and was taken to the USA by the Pilgrim Fathers; it was first standardized in America. Like the West of England, it is an auto-sexing breed. The Steinbacher was bred from the Chinese crossed with local strains of geese in eastern Germany. There are grey and blue varieties. The Sebastopol of eastern Europe is a light breed of exotic appearance, although it can be a productive layer and table bird. It is unique in having long, frizzled feathers, and it needs good management if the plumage is to be presented to its best advantage. There are two varieties recognized in Britain: the smooth, which has a smooth breast, and the frizzle that has curly feathers all over the body. In the USA, only the latter is recognized.

American: The American Buff is a heavy breed, similar in stance to the Embden but with attractive buff plumage. It has a good record as a productive utility breed in addition to its show features. The

Commercial Legarth strain of table geese. Photo: Early Bird Products

Chinese geese are distinguished by the prominent knob above the bill. They lay more eggs than other breeds of geese.

Prize-winning Sebastopol goose at the National Poultry Show. Its ballerina-type plumage makes it more appropriate for showing than for utility reasons.

A Brecon Buff goose with her family of young geese.

USA also recognizes the Egyptian, Tufted Roman and Canada breeds, but American breeders of the latter must have a federal permit before they can be sold.

Housing

Ventilation is the crucial feature when it comes to goose housing. Commercially, open-fronted sheds or pole-barns are used, so the geese are never far from the great outdoors. On a small scale, a shed, an adapted poultry house or indeed any movable shelter is suitable. There are also purpose-made goose houses for small numbers. Some breeders prefer to house breeding sets separately in individual runs, so that the 'bonding' or recognition of family grouping of gander and his mates occurs before the breeding season starts. They are then released to the main pasture. On grass, around twelve to fifteen breeding geese can be kept to each hectare (five to six per acre). A typical rotation is shown in the diagram

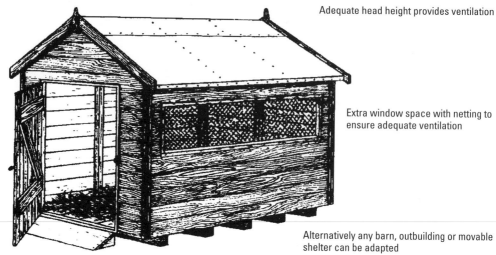

Adequate head height provides ventilation

Extra window space with netting to ensure adequate ventilation

Alternatively any barn, outbuilding or movable shelter can be adapted

Smooth ramp for ease of access

Adapting a shed for geese.

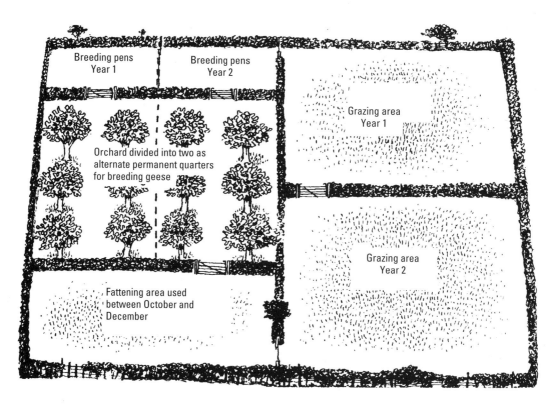

Breeding pens
Year 1

Breeding pens
Year 2

Grazing area
Year 1

Orchard divided into two as alternate permanent quarters for breeding geese

Grazing area
Year 2

Fattening area used between October and December

A convenient layout showing the rotation of land for geese.

Roman/Embden geese well camouflaged against the snow at Broad Leys.

on page 154 for areas specifically set aside for young growing geese before they are sold.

Our breeding geese were all-white Embden/ Roman crosses so they were reasonable in size. All the young were identical to the parents. They had field grazing where some crab apple trees grew, and the windfalls were always popular with them. Our birds were invariably supplied with a warm, straw-littered house, and just as invariably chose to ignore it. One particularly hard winter, when we had almost continuous blizzards, even finding them at feeding time became a major operation, so well camouflaged were they against the snowy fastness.

Feeding

The goose's main food, as referred to earlier, is good, clean, short-growing grasses. This can be supplemented with whole wheat, with the amounts varying depending on the season. More grain will be required when the grass has stopped growing, or is in limited supply.

Goslings do well on waterfowl starter crumbs and, as mentioned earlier, it is possible to buy starter and grower rations that are free of added medications. Goose starter feeds contain 20 per cent protein, grower rations have 18 per cent protein, and finisher rations are 16 per cent protein.

As they grow and are introduced to grass, the goslings can gradually go over to wheat instead of the starter ration. Where breeding is to take place, however, it is a good idea to feed the breeding stock a purpose-made breeder ration from winter to early spring. This has a higher protein content, as well as a range of minerals and vitamins to ensure that nutritional deficiencies do not reduce fertility or subsequent hatchability. Drinking water must be available at all times.

Breeding

From late autumn onwards is the best time to acquire breeding geese if you want to rear your own goslings in the spring. The usual mating ratio is one gander to every three geese for a heavy breed, and one to five for a light breed. If you buy an already established breeding set, there should be no problems, but it can happen that a gander (which in the wild is largely monogamous) will have nothing to do with a particular goose and will try to drive her away. This happened to us one year and we had no choice but to sell the unfortunate goose, hoping that she would eventually find a true paramour. (I am glad to say that she did.)

Once the breeding sets have been established, by keeping them separate from other groups during the late autumn to winter period, they can subsequently be released to join other sets all together in one field, and as long as the area is big enough, there will be no conflict because the ganders will respect each other's mates. But if a number of ganders and geese are turned haphazardly into a field, there will be fighting, and the resulting 'flock matings' will usually result in a considerable reduction in overall fertility. It is good practice to ring breeding geese, not only to differentiate them from younger stock, but also to allow performance records to be kept.

Many people have problems when it comes to breeding geese. One of the most common is the tendency for several geese to want to share the same nest, so that eggs in different stages of development are jumbled up. The only way to avoid this is to date each egg as you find it, and when the first goose sits, harden your heart and remove the second goose. You can try putting her in solitary confinement (possibly in a shed or small enclosure) with her eggs. She may decide to sit there. Sometimes a goose will choose an unsuitable site for a nest. One of our geese once nested in the open, under a crab apple tree. We

Optimum Conditions for Incubating and Hatching Geese		
	Temperature	*Humidity*
Chinese geese	Day 1–26: 37.5°C (100°F)	55%
	Day 27–31: 37°C (99°F)	75%
Light breeds	Day 1–26: 37.5°C (100°F)	45%
	Day 27–32: 37°C (99°F)	75%
Heavy breeds	Day 1–30: 37.5°C (100°F)	50%
	Day 31–36: 37°C (99°F)	75%

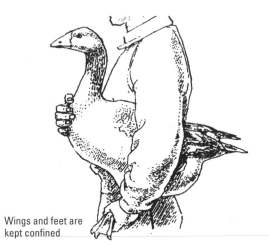

Wings and feet are
kept confined

Carrying a goose.

hastily put straw bales around her, laid a sheet of
galvanized metal on top, with an extra bale to weigh
it down. She approved, sat tight and eventually
hatched out all but one of the clutch.

Once the goslings have hatched, the main danger
comes from rats. Goslings must be protected against
them. A goose in the open with a group of goslings
cannot defend them, for rats are capable of taking
one without her realizing.

Weather protection is also required, particularly in
the first six weeks, until they are hardy and fully
feathered, although it is important that they go out as
soon as possible during the day. We once had a
gosling that managed to get out of a small run and
was nowhere to be found when its brothers and
sisters were brought into the conservatory to sleep in
a jumbo-sized cardboard box overnight. Later, when
David took Sandy the dog for his last walk up the
lane, he found a pathetic white blob on the grass
verge. Yes, I know one is not supposed to give
hypothermia victims alcohol, but I administered
some brandy and warm water via an eye-dropper and
put it in a warm place overnight. Next morning the
gosling was fully recovered and ready to go out
again.

It is important not to let the goslings get wet in the
brooder. If they do, they will start pecking at each
other's plumage. Avoid open drinkers. A traditional
remedy was to cut a turf of fine, clean grass so that
they give this their attention instead.

Geese that are raised for the table are usually to
cater for the Christmas market, or in lesser numbers
to be sold as 'green' (young) geese for the
Michaelmas period in Britain. This is in late
September. Killing must only be done by an
experienced person. Plucking and eviscerating are as
detailed for table chickens.

Dealing with Problems

Geese are extremely hardy, although they can suffer
from the problems that were outlined for ducks. If
managed well, with good food and fresh, clean
pasture and water, they are far less likely to succumb
to infections. There are some specific conditions to
watch out for.

Gizzard Worm

This parasitic worm can be a menace where young
geese are concerned. Traditionally, the condition
was called 'going light'. Adults can tolerate a certain
level of infestation, but the only way to avoid it is to
follow a rigid pattern of grazing rotation, ensuring
that the young geese have the fresh grass. The
warning signs are a bird that sits down a lot and is

*Goose nesting in a garden shed. Note the size of the nest
she has made.*

reluctant to move when approached. The eyes are still and slightly staring, a symptom difficult to describe but once experienced easily spotted. When the young goose is at this stage it requires immediate action or it will die. A poultry vermifuge such as Flubenvet can be added to the feed.

Slipped Wing

Also called 'angel wing', this is a condition where the wing sticks out at an angle, with the bird being unable to place it at rest, against the side of the body. There is nothing that can be done to cure the condition in an adult. In a gosling, it may be the result of over-feeding on protein-rich food. Give a grower ration with a reduced protein content, and then strap the wing in place – soft, wide, cotton-covered elastic works well. Check if there is any improvement after a week. If caught early enough, the condition can sometimes be cured. It is best not to breed from such a bird, for it is still not clear whether or not there is an inherited tendency towards it.

Fallen Tongue

This is a condition only seen in geese with dewlaps under the chin. The tongue literally falls down into the dewlap and becomes lodged there. The only way to deal with it is to physically lift the tongue back, then tie off the bottom of the dewlap in order to make it smaller. It is unsightly, but enables the bird to eat and drink again. The alternative is to cull the bird.

Aggressive Gander

Ganders can be fierce, especially in the breeding season. To tame him, make your hand into a goose-head shape and move it up and down in front of him. This is an aggressive sign, and if he has any sense he will give way. Unfortunately, there are some ganders that are devoid of sense. If he still advances, making a circular motion to grab him by the neck will throw him off balance, but beware of the powerful wings. If these techniques fail, the best protection is to arm yourself with a dustbin-lid shield when entering the goose enclosure.

While goslings still have their baby down plumage they are vulnerable to hypothermia and need night-time protection, and protection against bad weather.

Commercial goslings in protected conditions. Once fully feathered they are ready to go out on grass. Photo: Early Bird Products

Christmas geese. Photo: John Adlard

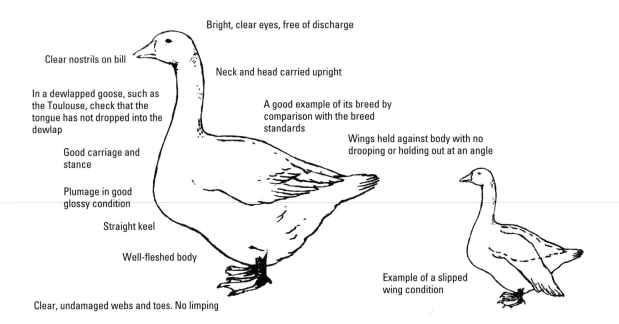

Bright, clear eyes, free of discharge

Clear nostrils on bill

Neck and head carried upright

In a dewlapped goose, such as the Toulouse, check that the tongue has not dropped into the dewlap

A good example of its breed by comparison with the breed standards

Wings held against body with no drooping or holding out at an angle

Good carriage and stance

Plumage in good glossy condition

Straight keel

Well-fleshed body

Example of a slipped wing condition

Clear, undamaged webs and toes. No limping

What to look for in a healthy goose.

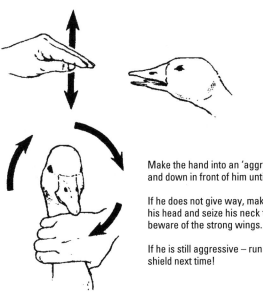

Make the hand into an 'aggressive gander' head and move it up and down in front of him until he backs away.

If he does not give way, make a quick circular motion in front of his head and seize his neck throwing him off balance, but beware of the strong wings.

If he is still aggressive – run and arm yourself with a dustbin-lid shield next time!

Coping with an aggressive gander.

13 Turkeys, Guinea Fowl, Quail, Peacocks and Pheasants

Turkeys

Beef, mutton, and port, shred pies of the best, pig, veal, goose and capon, and turkey well drest. (Tusser, 1573)

Central America in general and Mexico in particular are the most likely areas where turkeys originated, for it is known from Aztec artefacts that this civilization had domesticated them. They were introduced to Spain by the Conquistadors, with 1524 being the first reliably authenticated date. From Spain, they quickly spread to other areas of Europe and the East. The Pilgrim Fathers and other early settlers in America took turkeys with them from Europe. These then crossed with the native wild turkeys found in the wooded areas of the new world. Some of these hybrids were, in turn, brought back to England, where a gamekeeper called John Bull began to select them for a broader breast. He subsequently emigrated to Canada, taking some of his stock with him, and then sold some of the broad-breasted birds to the USA.

Breeds

In addition to the five recognized sub-species of wild or bush turkey, there are several breeds that have been domesticated and developed. Similar breeds often have different names and standards in different countries, and commercial breeds have also been hybridized for quick growth and size. Some strains are often too big for natural mating to take place, and artificial insemination is required. Those with an interest in breeds will prefer more 'undeveloped' strains, as will small farmers catering for those with normal ovens!

Black (Black Norfolk): In Britain, the Black Norfolk is the only recognized black turkey, the name indicating the area of the UK where it was found in greatest numbers. No doubt from the same line as the Spanish 'Negra de Pavos', it was reintroduced by the early settlers to America.

In the USA, the Black is the recognized breed. Like its British cousin, it is required to have plumage that is dense all-over black. However, the head and face can be red, changeable to bluish-white, while British standards require it to be as red as possible. The American, Spanish and French Blacks tend to have more lustrous plumage, while in Britain it is a duller black.

Bronze: This is the closest in colouring to the indigenous wild turkey of the USA, and is an overall dark bronze. Edward Brown claims that the Cambridge Bronze was developed from the Black in Cambridge – whose turkeys, according to the Rev. E.S. Dixon in 1850, had: 'Plumage which varies very much; sometimes made up of shades of reddish-brown and grey.' Crossing these with the American Bronze, according to Brown, produced the Cambridge Bronze.

There are three types recognized by the British Poultry Standards: the Bronze, the Cambridge Bronze, and the Black-Winged Bronze or Crimson Dawn. The first has a metallic sheen to the plumage, black-and-white barred flight feathers, and a black, brown and white-edged tail. The second is duller in hue, and the body feathers are tipped with grey or white. The third is the same as the Bronze except for the primary feathers, which are all black.

The American Standard of Perfection recognizes only one breed: the Bronze that has no white in the body plumage, but has black-and-white barring on the primaries, and brown on the back and tail.

Commercial interests in the USA developed the Broad-Breasted Bronze, possibly based on John Bull's past introductions, referred to earlier. These in turn came back to Britain as one of the best-selling commercial breeds, until ousted by commercial development of the White in this century.

White: Originally a mutation of the Black, the White seems to have been developed in Europe; from here it was introduced to the USA as the White Holland, indicating its place of origin. There are also past references to an 'Austrian White'. In Britain, the only recognized white turkey is the British White. In the USA it is still called the White

An adjustable height feeder being used by some Bronze turkeys. This type of feeder prevents wastage of food and loss to wild birds. Photo: Hengrave Feeders

Holland, though commercial development there has also hybridized it to produce the Broad or Large White. In 1951, the Beltsville Small White was recognized by the American Standard of Perfection; this was initially developed as a small commercial breed by the US Department of Agriculture at Beltsville. It met the need for a small, compact turkey for those who did not want massive-sized birds; it is also available in the UK.

Commercial development of the White has produced Maxi, Midi and Mini strains to accommodate producer preferences; but as already mentioned, very large birds are incapable of mating naturally.

Buff: Originally called the Fawn in Britain, it is now recognized as the Buff. Edward Brown was of the view that it was a mutation of the Bronze when crossed with White. It is a beautiful cinnamon brown with white primary and secondary feathers, as well as white tail feathering.

Bourbon Red: Brownish-red in colour, the flight feathers of this breed are white, while the tail is white with a thin bar of brown near the bottom of the tail. It is rare in Britain and, together with the Bronze, Slate, and Spotted or Pied breeds, comes under the auspices of the Rare Breed Survival Trust's Turkey Project.

Slate (Blue): Recognized as the Slate in the USA, this breed is called Slate or Blue in the UK. The plumage is an overall slate-blue colour, as the name implies. Edward Brown called it Grey, pointing out that it was common in Cambridge and Norfolk where it was called the Bustard turkey. He also referred to its high quality as a table bird in Ireland, and expressed the wish that they would standardize it there to produce an Erin breed. Alas, it never happened. There is also a Lavender breed.

Black and White: There are a number of breeds with black and white feathering, with most being found in the USA. The Narragansett, bred in the Narragansett Bay area of New England, is thought to be the product of wild turkeys crossed with Black turkeys brought from Europe by the settlers. The patterning is similar to the Bronze, except that the bronze colour is replaced by steel grey with black-and-white barring.

The Royal Palm is pure white with black banding; in Britain, the same type is called the Cröllwitzer. There is also a Spotted or Pied variety in Britain, a mutation that appeared in 1947 in a closed flock of Broad-Breasted Bronze.

Housing

Traditionally, turkeys were raised outdoors on sandy soil where the drainage was efficient and therefore a protection against blackhead protozoans becoming established. Well ventilated pole-barns or sheds are ideal. If the overall site is secure against foxes, the barn can be open-fronted, facing away from prevailing winds. Alternatively, a wire-meshed wall

<table>
<tr><th colspan="1">Flock Density for Free-Range Turkeys</th></tr>
</table>

Flock Density for Free-Range Turkeys

Maximum stocking rate in fixed houses – 2 birds per
  square metre
Maximum stocking rate in mobile houses – 3 birds per
  square metre
Maximum outdoor stocking rate – 800 birds per hectare

Cröllwitzer turkey. Note the perching platforms in the background.

and door can be used to secure them at night.

Wood shavings or chopped straw to a depth of 7.5cm (3in) provides an efficient absorbent layer for droppings. Perches are popular with the birds, as long as they are not too high. Straw bales are also suitable for this.

Suspended feeders and drinkers are best because litter is not scratched into them. They can also be progressively raised as the turkeys grow.

It is important not to panic turkeys. They have a tendency to flap, and once one of them starts, the rest join in, and before you know where you are, there is pandemonium. The danger here is that they will hurt themselves, either bruising each other or even drawing blood by banging their wings against the walls. The golden rule is to approach turkeys in a quiet, calm way, with no sudden movements. They do get used to the same person feeding them.

Feeding

Turkeys need a high protein diet, and there are natural proprietary rations for turkeys, without all the additives that are in feeds for the intensive sector. They include a starter ration with 25 per cent protein from hatch until seven weeks; a grower's ration with 21 per cent protein until two weeks before killing; and a finishing ration with 18 per cent protein. All these are available as natural feeds, without in-feed medication. Alternatively, turkeys can be given mixed grain that includes maize and soya, with grit and oystershell from around twelve weeks onwards. Giving them access to grass ranges will also allow them to glean some of their own food. On a small scale, they can be fed morning and evening, although some producers feed turkeys three times a day. Fresh water is essential at all times. It is a good idea to have the drinker placed above a metal grille so that any spilled water can drain away, without making areas of litter damp. This will reduce the chances of disease organisms such as coccidiosis and blackhead proliferating.

A small turkey house made of recycled materials at the author's smallholding. It is south facing and has plenty of ventilation. The turkeys have access to pasture during the day.

Optimum Conditions for Incubating and Hatching Turkeys		
	Day 1–24	Day 25–28
Temperature:	37.5°C (100°F)	37°C (99°F)
Relative humidity:	55%	75%

Free-ranging white turkeys being raised for Thanksgiving in New England. Note how the tree trunk has been protected against bark stripping.

Breeding

Male turkeys are called stags or toms. The usual mating ratio is around ten females to one male, although this can vary, depending on whether the birds are light or heavy breeds. Reference has already been made to the fact that some large hybrids are incapable of natural mating, and artificial insemination must be used.

Turkey eggs are thicker shelled and more pointed than those of chickens, so can be difficult to candle if they are being artificially incubated. Hen turkeys will incubate and hatch their eggs, if conditions are suitable. A larger-than-normal sized nest is needed, as well as a quiet place where she will not be disturbed. We found that a corner of the barn with appropriately arranged hay bales always met with approval. The average incubation period for turkeys is twenty-eight days.

Similar brooding conditions to those provided for chicks or ducklings are suitable. Where day-old or young turkey poults are bought, they are often supplied as A/H stock, meaning 'as hatched'. In other words, they will be a mixture of males and females. The males grow more quickly than the females, and also grow bigger. Chick crumbs or a turkey starter ration can be fed first. Some turkey rations include an anti-coccidiosis and anti-blackhead medication. These are usually utilized where rearing is on a large scale. On a small scale, where good management prevails and damp areas are avoided, there is less likelihood of a problem. Traditionally, turkeys were never allowed onto ground that had been grazed by chickens, so the risk of blackhead infection was reduced. In practice, if pasture is rotated sensibly, if the management is good and there is ample space for all the poultry, turkeys and chickens can live perfectly well together.

Where turkeys are being raised for the Christmas or Thanksgiving markets, leave enough time to kill, pluck and draw the birds before hanging them in a cool room. The birds should not be fed for at least twelve hours before killing, although water should be available. Plucking and gutting are as detailed in the section on table chickens, although the leg tendons are much stronger and will need special attention to remove them. Welfare, slaughtering, food safety and labelling regulations will apply, and the premises must be registered.

Coping with Problems

Turkeys are extremely hardy and can tolerate a considerable degree of cold, but they dislike wet and windy conditions. As with other poultry, a good level of ventilation in their house will go a long way to avoiding many of the problems associated with intensive rearing.

Parasitic worms can be picked up, although effective grass rotation will minimize this; Flubenvet added to the feed is effective for all poultry. Turkeys also need to be checked for external parasites, and treated with a proprietary product such as a pyrethrum-based powder, if necessary.

The main diseases associated with turkeys are blackhead (*Histomoniasis*) and coccidiosis, both caused by protozoans that attack the digestive system. They can be prevented by the use of feeds dosed with antibiotics, though most people would prefer not to use these. The best approach is in providing excellent pasture and management conditions so that damp areas of ground, long wet grass and damp litter are avoided.

Guinea Fowl

The African fowls are big, speckled, humpbacked, and are called 'Meleagrides' by the Greeks.

(Varro. 40BC)

The west coast of Africa is thought to be the place where guinea fowl originated, and they are still

The author's six-week-old turkey poults. They are ready to go out on pasture.

found in the wild around the Guinea coast. They were considered to be a delicacy in the courts of Europe 500 years ago, and bore the common name 'Tudor Turkeys'. According to Lewis Wright in his *Book of Poultry,* guinea fowl were often to be seen perched in the rigging of sailing ships docking at Bristol from West Africa. The slave trade probably also introduced them to Jamaica and the Cape Verde Islands.

At first sight, the guinea fowl looks completely out of proportion, with its round body and tiny head, but it can run at a fair pace and is adept at flying up to high perches. The sexes are identical in size and feather colouring, and are difficult to differentiate; when adult, both weigh 1.3–1.8kg (3–4lb) live weight. They are very hardy birds. Their tight feathering provides reasonable protection against the rain, and they adapt well to relatively cold temperatures. The guinea fowl's sharp eyesight also serves it well when it comes to spotting insects: it devours these with relish. It can be used to good effect in clearing a site of particular pests such as weevils, beetles and slugs, though it does not distinguish between these and the useful ones. When I was in California some years ago, I saw these fowl being used in potato fields to control Colorado beetles.

They are vocal birds, a factor that makes them useful as guard dogs but unpopular with close neighbours. The male utters a single shriek, while the female has a two-syllable call similar to 'come back, come back' (in America, they claim that she says 'buckwheat, buckwheat'). If suddenly alarmed, the female will also utter a single note shriek, but the male never uses a two-note call.

Breeds

There are no breed standards for guinea fowl as there are for poultry in Britain and the USA. Where they are displayed at shows, the emphasis is on nicely coloured and evenly marked birds. For commercial table use, some selective breeding of the Pearl has taken place, particularly in France, Holland and Italy.

Pearl: This is the most common variety. It has dark, purplish-grey plumage covered with white dots (pearls). In the wild this provides effective camouflage against the dappling effect of tree foliage.

White: Originally a mutation or 'sport' of the Pearl, this is pure white. The skin is also lighter than that of the Pearl.

Lavender: The markings resemble those of the Pearl, but the plumage is a much lighter, dove grey.

Other mutations: Other mutations and crossings

have produced a range of colour variations and 'splashings', some rare. Markings may also be more striped than spotted in some strains. In addition to the Pearl, White and Lavender, the following varieties are available in the UK:

Pied (Pearl/White): The original grey with white on the sides and wing tips.
Clear Lavender: As the Lavender, but minus the spotting or stripes.
Lavender/Pied: As the Lavender, but with white on the sides and wing tips.
Mulberry: A purplish body and buff head.
Mulberry/Pied: As the Mulberry, but with white sides and wing tips.

Colour varieties available in the USA include purple, blue, violet, chocolate, brown, buff, porcelain, slate, chamois, fauve, pied and silver-wing.

Housing

Those who keep guinea fowl in small numbers often find that they do not even require housing because they prefer to roost outside, often in trees. However, in areas where winter weather conditions can become severe, they should always be offered the option of a barn or other shelter. It is also important to bear in mind that a fox will take guinea fowl, although their sharp eyesight is an excellent defence. It is said that guinea fowl will not only spot a fox or hawk before they can get in range, but will also warn other, more vulnerable birds of their approach.

On a larger scale, any well ventilated outbuilding, barn, shed or poultry house can be adapted for them.

Guinea fowl make excellent 'guard dogs', though their strident calls may not be popular with neighbours.

On a commercial scale, it is not a good idea to have more than 100 birds together. Guinea fowl are easily panicked, and where they are kept in large numbers, can crowd into one end of a house, leading to possible suffocation. Stretching 90cm (3ft) high wire netting across the corners of a building helps to prevent this happening. Buildings with open-meshed walls such as those used for turkeys are ideal, particularly if they have access to open-air pens as well. The outside pen fencing needs to be at least 1.8m (6ft) high, otherwise they will be able to fly over.

Feeding

Guinea fowl chicks (keets) do well on chick crumbs for the first few weeks, before gradually going over to an additive-free adult ration. Some commercial grower's rations produced for the intensive sector are toxic to guinea fowl, so free-range and organic feeds are best. A free-range chicken layer's ration is suitable.

Where a wide free range is available, and insects provide a proportion of the protein requirements, mixed grain only can be given; however, this needs to be fairly fine because the guinea's crop is small. Grain needs to be supplemented with grit and crushed oystershell, unless a natural free range provides these elements.

For a small number of 'pet' guinea fowl, a 'wild bird mixture' available at pet shops and garden centres is suitable. Clean water should be available at all times.

Guinea fowl for the table are available as fertile eggs or as day-olds. They are reared until they are between 11–14 weeks old, when the deadweight in feather will be upward of 1.2kg (2.75lb). As oven-ready birds these will be approximately 0.9kg (2lb), but these figures do depend on a number of factors, including strain of bird, housing, management and feeding. Plucking and eviscerating are as detailed for table chickens.

Breeding

In the wild, guinea fowl form breeding pairs, and this pattern occurs also in domestic flocks if there are similar numbers of males and females. Most people, however, will tend to have more females than males, in the ratio of four or five to one. It can be difficult to distinguish between the sexes, although the differences in calling will be apparent from around eight weeks of age. Other points of contrast are the larger 'helmet' and wattles of the male, and the generally 'finer' head of the female.

Free-range guinea hens usually come into lay in spring, and will go off to lay their eggs in clumps of

Optimum Conditions for Incubating and Hatching Guinea Fowl		
	Day 1–25	*Day 25–28*
Temperature	37.5°C (100°F)	37°C (99°F)
Humidity	55%	75%

long grasses, in shrubbery, or underneath the hedge. Normally they lay between twenty and thirty eggs before becoming broody. The incubation period is twenty-eight days, the eggs 'pipping' at around twenty-five days.

The eggs are thick shelled, and candling them to check on the contents is difficult. Broody chickens and turkey hens have been used successfully to incubate guinea fowl eggs: a chicken the size of a Plymouth Rock can take up to fourteen, while a turkey can take up to twenty-four. Crosses between chickens and guinea fowl are not unknown, but the progeny are invariably sterile.

If eggs are removed regularly so that broodiness is prevented, a guinea hen lays between thirty and 100 eggs a year, depending on the strain. The eggs can, of course, be eaten; the yolk is larger in relation to the albumen than is the case with chicken eggs.

The young keets are pretty little things, rather like baby quail. If hatched artificially, they will need protected brooder conditions, with a lamp until they are fully feathered. At six to eight weeks they are hardy enough to go outside. A broody hen will take care of her charges, and it is interesting that the young guinea fowl will tend to look to her even after she has lost interest in them. This can be useful as a means of control, if you are trying to get them inside.

Dealing with Problems

Guinea fowl can suffer from any of the ailments that affect chickens, but they are extremely hardy and on a small, extensive scale there are unlikely to be problems. It is wise to check for external parasites, and also for signs of worm infestation, when the bird will be thin and the feathers will look rough. A good policy is to check them and treat them with a pyrethrum-based insecticide, or other suitable pro-prietary product, at the same time that you treat all the other poultry. Similarly, all the birds can be wormed once a year in late summer with a vermifuge such as Flubenvet, administered in the food.

Quail

We loathe our manna and we long for quails.
(Dryden, 1682)

The quail was known to all the great civilizations of the past, and even appears as one of the letters in ancient Egyptian hieroglyphics! Quail are classified as game birds and there are more than forty species, but only a few are kept in captivity; these are mainly Coturnix, Bobwhite and Chinese Painted (referred to as Button quail in the USA).

Coturnix Quail

No one can be certain how the development of the Coturnix laying quail took place, but it is based on three types: the wild Coturnix quail, *Coturnix coturnix*; the Eurasian or Pharoah quail, *Coturnix communis*; and the Japanese quail, *Coturnix japonica*. In recent years, laying quail have been developed for commercial purposes, and if eggs are required in any quantity, or table birds are to be raised, these are the strains to choose: a commercial strain can produce up to 300 eggs a year under the right conditions. With their brown and cream speckled shells, they are about a third the size of a chicken's egg, but are regarded as a quality, delicatessen product.

Males and females are virtually identical, although the hen quail is slightly larger, at an average length of 19cm (7.5in). Both have dappled, dark brown, buff and cream, striated backs, paler underbellies, breasts and flanks. The females are slightly paler, and the males have a more reddish breast, and it is only when this feature becomes detectable at around three weeks of age that it is possible to sex them. There is a distinctive stripe above the eye and a white collar, although these may be less apparent in the female. There are coloured varieties of Coturnix available, and you may prefer to keep some of these. They are good egg layers, although not quite as productive as the commercial strains, but fine for household use. We have kept most of the coloured varieties, and would be hard put to decide which are the favourites.

Fawn: The Fawn variety is a lovely pinkish brown with the feathers pencilled in white. The eye stripes are present, although not as distinctive as in the plain Coturnix.

English White: Good specimens are pure white with just a hint of eyebrow lines, but most tend to have a few black markings.

Female Coturnix quail. Note how the head feathers have been pulled by the mating activities of the male.

Tuxedo: Here, dark brown and white markings are fine, as long as they are where they should be! The bird is so called because it has a dark brown back and a pure white breast, rather like a waistcoat or tuxedo.

Range: This variety is confusingly called American Range in Britain, and British Range in the USA, proving George Bernard Shaw's claim that we are 'two nations divided by a common language'. Some call it the Brown quail because of its all-over dark brown plumage, but this is the name generally accorded the Australian Brown quail.

Manchurian (Golden): This is essentially the same as other Coturnix, with similar markings, but having a beautiful golden hue to its feathers.

Bobwhite Quail

The Bobwhite, *Colinus virginianus*, is much bigger and is primarily regarded as a game bird in the USA, although the Eastern variety, that is slightly heavier, has also been raised for eggs and meat. In Britain, the Bobwhite is regarded as an aviary bird. The male is around 25cm (9.5in) in length, while the female is slightly larger. The back, tail and brow of both sexes are dark brown, while the chest, belly and flanks are lighter. A white stripe covers the eyebrows, and in the male there is a white bib under the chin.

The Bobwhite is a perching bird, and must be provided with a perching area. A tall aviary with an outside run is ideal, particularly where shrubs and branches are available.

Chinese Painted (Button) Quail

The Chinese Painted (Button) Quail, *Coturnix*

chinensis, is popular in aviaries or as a pet. In size, it is around 12cm (4.5in) long, making it the smallest species of quail. The female is slightly larger. The males are aggressive to each other and must be kept apart or they may fight to the death. A male and female form a monogamous pair.

In appearance, the breed is compact and round. The male has a brown- and black-flecked back and crown, and bluish-grey breast and face. The belly and tail feathers are reddish brown, while the chin and throat have distinct black and white striping. The female is an overall mottled brownish hue from the fine black and white flecks. Her back is slightly darker than her abdomen. She has a white patch on the throat but no barring. In both sexes, the beak is black and the eyes are brown.

They are ground dwellers and cohabit well with flight birds in an aviary. Suitable companions include finches, canaries and birds of similar size. Larger flight birds may attack them.

There are around ten sub-species, and no one knows precisely what the origins are of many of the birds kept and bred in captivity. It is likely that they have been bred from various sub-species. In recent years there has also been considerable breeding for colour variation. They include:

Silver: This is the most common mutation and has all the feathers in varying shades of light grey.

White: These are all-white, although not albinos. There may be a few coloured feathers in some birds, but the aim is to produce a snow-white effect.

Red-Breasted: The face is almost black with a fine white line around the throat. The red area extends from the vent, across the breast to the throat.

Red-Breasted Silver: Here the pastel grey plumage contrasts beautifully with the pinkish-red breast.

Fawn or Cinnamon: Originally developed in Australia, this is called Fawn in Europe and Cinnamon in the USA. There is also a Blue-Faced Fawn and a Red-Breasted Fawn.

Blue: The plumage is a dark, overall blue.

Black: Bred by an American breeder, this is an even darker blue, approaching black.

Ivory: Lighter than silver, these have an overall ivory hue, although the male's breast is grey.

Fawn Coturnix quail

Bobwhite quail in an outside aviary.

Chinese Painted or Button Quail in a grassed aviary.

Golden Pearl: Originating in Europe, this has yellow feathering with light brown barring. There is also a Fawn or Cinnamon Pearl, and a Blue-Faced Pearl.

Housing

Outdoor aviaries are ideal for all types of quail, although they will need a protected sleeping area, for they are not hardy. The optimum conditions are a temperature of 16–23°C (62–73°F), and a relative humidity within the range of 30–80 per cent. An aviary is essential for the perching Bobwhites. Laying Coturnix are more ground-orientated, although they do sometimes fly up, and may knock their heads on the roof! It may be necessary to place some netting just below the roof as a safety net. Chinese Painted Quail are the most ground-orientated of all, scuttling about like little mice. Any house that is dry, well ventilated yet free of draughts is suitable for them – though most importantly, it must be secure from rats! What the fox is to chickens, the rat is to quail.

Commercially, most quail are kept in cages, although some producers do keep them as floor-reared stock. Here, wood shavings or chopped straw are used as litter. Those who are interested in keeping them commercially, but who dislike the idea of wire-floored cages, could consider the litter-floored cages that are also available. In these they are at least able to scratch about.

Our system of housing was to use rabbit hutches for our Coturnix layers. These were adapted by having the roof of the daytime area removed and replaced with wire mesh. The sleeping area was left covered and this is where they also laid their eggs. A small light was used to extend the natural daylight in winter, so that egg production did not decline too much. In summer, or at other times when the weather was mild, they were housed in moveable runs on the grass outside. In winter the non-utility Coturnix and the Chinese Painted Quail were in canary-breeding cages in the conservatory, while the Bobwhites were in a parrot-breeding cage with a perch. We would often let them out for exercise in the conservatory, and they never showed signs of being aggressive to each other. This is probably because they were all tame and used to being handled. In summer these quail were also allowed to range on grass.

Feeding

Quail have small appetites compared with chickens, although their protein requirements are quite high. They also need food that is composed of smaller

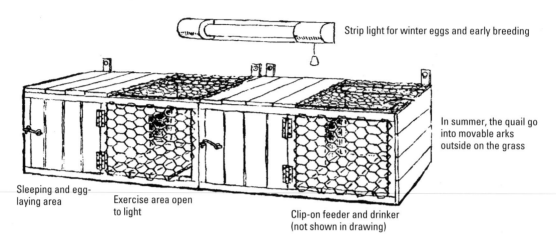

Strip light for winter eggs and early breeding

In summer, the quail go into movable arks outside on the grass

Sleeping and egg-laying area

Exercise area open to light

Clip-on feeder and drinker (not shown in drawing)

Author's system of adapting rabbit hutches as winter housing for quail.

particles than the average layer's pellets mixture. Natural chick crumbs without in-feed medication is suitable for young quail, as well as Chinese Painteds. There are also commercial quail layer's pellets available. These have a protein level of 20%. In addition to concentrates, quail will need access to fresh water at all times.

On a non-commercial basis, canary seed and millet can be given. Bunches of greens to peck are also popular. These include lettuce, chickweed and parsley. As with poultry, limestone and fine grit should be available to them when they need it for the proper digestion of grain, and also to ensure strong eggshells. As quail have sharp beaks, it is a good idea to provide a piece of cuttlefish bone from the pet shop, as one would for aviary birds. If Chinese Painted Quail are kept with finches in an aviary, their gleaning of fallen seeds should not be regarded as sufficient diet. There must be a basic ration of a proprietary feed given as well.

Breeding

Quail will breed in captivity, but it is often difficult to get them to brood their own eggs; aviary conditions where there are shrubs and long clumps of grasses for secret nests are more conducive to this. Generally the eggs are incubated artificially.

Most incubators are suitable, but a higher hatch rate will be obtained if the eggs are dipped in warm water to which an egg sanitant has been added, before they are put in the incubator. An automatically turned incubator is essential.

Newly hatched quail are about the size of bumble bees, and they can easily get lost under normal brooding conditions. My method of brooding was to use an old fish tank with netting placed over the top, and a heat lamp above; there were no chinks for them to get out of, and they were quite safe in there. Poultry sawdust can be placed in the bottom as litter, but I found that if paper towels were put on top before they went in, this prevented them eating litter, which can cause impaction of the digestive system.

Optimum Conditions for Incubating and Hatching Quail

Coturnix layers	*Bobwhite*	*Chinese Painted*
Day 1–14: 37.5°C, humidity: 45%	Day 1–19: 37.5°C, humidity: 60%	Day 1 – Day 11: 37.5°C; humidity: 45%
Day 15–18: 37°C, humidity: 75%	Day 20–23: 37°C, humidity: 75% (Note: 37.5°C is 100°F and 37°C is 99°F)	Day 12–16: 37°C, humidity: 75%

Relative egg sizes, from left to right: A small chicken egg, Coturnix quail, and Chinese Painted (Button) quail.

One of the author's newly hatched Coturnix quail chicks.

Eggs of Coturnix quail in a purpose-made small carton.

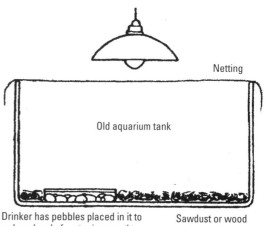

Netting

Old aquarium tank

Drinker has pebbles placed in it to reduce level of water in case the baby quail drown

Sawdust or wood shavings litter

A safe brooder for baby quail.

Once they have learned to feed normally, the towels can be removed. A drinker should have clean pebbles placed in it otherwise the tiny chicks may drown.

As is so often the case, there are far fewer problems with small-scale, non-intensive practices than there are with intensive farming. Although quail can be affected by most of the poultry diseases, these are unlikely to appear if cleanliness, good management and feeding are given the necessary attention.

Quail Eggs

Quail eggs, both fresh and pickled, are popular. It is possible to buy egg boxes appropriate to the size of the small eggs. Hard-boiled eggs left in their shells can be pickled in brine and packed in small glass jars that will display the shells. The hard-boiled eggs are placed in the previously sterilized jars, and then covered with a brine solution made up of 50g salt to 600ml water (2oz per pint).

Hard-boiled eggs that have had the shells removed can be pickled in vinegar. The eggs should not be too fresh otherwise they are difficult to peel. Boil them, and then plunge them into cold water to make them easier to peel. The pickling solution that I use is made up of 1l (1.8pt) of white wine or cider vinegar, with 0.5l (0.9pt) water. Add two level dessertspoons of salt, a medium-sized chopped onion, and a sachet of pickling spice. Heat in a stainless-steel saucepan, and simmer for a couple of minutes. When it has cooled down, strain the liquid and pour over the shelled eggs in a jar. A few peppercorns and a chilli pepper from the pickling sachet can be added to make the whole jar look more interesting.

Miniature Scotch eggs are also attractive: the hard-boiled eggs are shelled and then dipped in flour before being coated with sausage meat. The eggs are then given a final coating of breadcrumbs before being deep-fried.

Peafowl and Ornamental Pheasants

Once in three years came the navy of Tharshish, bringing gold and silver, ivory and apes, and peacocks.   (Old Testament)

Peafowl are stately birds, at their happiest where they have plenty of space to roam; but their harsh calls preclude them from areas where neighbours are likely to complain. They are members of the pheasant family, and indeed owners of peafowl often breed ornamental pheasants as well, for their management and feeding are similar. The Blue (Blue Indian) is the most common species: it has a glossy purplish-blue head with crest, while the long train-feathers are decorated with ocelli (eyes). There are several varieties of the Blue Indian. The most common of these are the Black Shouldered, the White, the Silver and Pied.

The main commercial value in keeping peafowl is in selling fertile eggs for hatching, peachicks for rearing, or adults for breeding. There is also a small, but ongoing market for peacock feathers: the train-feathers moult naturally, and it is just a question of picking them up. Ornamental pheasants are usually sold to those stocking outside aviaries.

A house should be provided, although the birds may be reluctant to use it in all but the severest weather. Requirements are relatively simple: a height sufficient for a perch to cater for the male's long train, and sufficient protection and ventilation to ensure good health. A flat perch 5 × 10cm (2 × 4in) placed 1.8m (6ft) above the ground will allow the male's train to hang down in an unrestricted way. It may be a good idea, however, to have stepped perches, too, say at 60cm (2ft) and 120cm (4ft), so they do not have to jump down from a considerable height, thus risking foot injuries.

A trio of one male to two females is common, but if more are kept, the ratio should not exceed five

White peafowl at home in the spacious and elegant surroundings that suit them best.

The Blue Indian peacock is the most common breed.

females to one male. Two males will fight, especially in the breeding season. A hen will be fertile in her first year, but is best mated to an older male. Peahens are not renowned for being good mothers, although if eggs are left in a clutch, they will often sit and hatch them. Around thirty eggs a season are laid, but this number can be increased by taking them away as they are laid. Eggs are cream and glossy, and about the size of a turkey egg.

The chicks will need protected and heated brooder conditions until they are fully feathered. A chick crumb ration with a coccidiostat is advisable for the first six weeks, and after that, a free-range layer's ration is suitable, particularly if there is good ranging available. Peafowl are scratchers and foragers, and will glean a certain proportion of their proteins and other nutrients from insects. Access to clean drinking water is essential during daylight hours. Greens such as blown lettuce, dandelions and chickweed are popular. As with other birds, they should be checked for external parasites, and treated accordingly. They may also need worming once a year, depending on the condition of the ground.

Ornamental pheasants require similar conditions to peacocks, and do well as aviary birds.

Optimum Conditions for Hatching Peafowl and Ornamental Pheasants

	Day 1–25	Day 25–28
Temperature:	37.5°C (100°F)	37.0°C (99°F)
Humidity:	50%	75%

Note: Different breeds of ornamental pheasants vary in incubation times from 23–28 days. Temperature and humidity will therefore need to be adjusted accordingly, at 20–25 days.

14 Pigs

They are great softeners of the temper, and promoters of domestic harmony.

(William Cobbett, 1822)

The pig is descended from the wild boar, *Sus scrofa*, that was widely hunted all over Europe. It was also domesticated at an early stage. In the Domesday Book, the value of woodland was calculated on the basis of how many hogs it would support. With the increasing emphasis on the need for more woodlands to counter the effects of global warming, and the general reaction against intensive farming, it is interesting to note that the concept of using woodlands for livestock is rapidly gaining credence again; in Britain it is one of the prime areas of research within the field of agriculture. Small-scale breeders have always known that the best way of rearing any creature is to give it the conditions to which it is innately adapted, but it is always satisfying when science catches up with common sense.

Pigs are highly intelligent and sensitive creatures; it has been said that they are the most intelligent livestock on the farm. They are also naturally clean animals, and will deposit dung in just one area. As long as they are given sound, weatherproof and draught-free housing, clean conditions, adequate food and water and healthy exercise, they will do well. To keep pigs in Britain and the USA, it is necessary to be registered with the relevant authorities, and every animal must be suitably marked, tattooed or tagged for identification; also, practices must conform to animal welfare legislation. There are essentially several areas of activity that are relevant to the small-scale producer: raising weaners for slaughter; breeding pedigree animals; raising and selling selected cross-breeds; and keeping pigs as pets.

Raising weaners: Young, weaned piglets, or older store pigs, are bred or bought and raised until they are ready for slaughter. This may be to provide pork or bacon for the household, or to offer for sale locally or on the open market. Catering for the organic meat sector is the best option if the aim is to sell them.

Pork pigs may be lightweight, weighing 45–68kg (100–150lb); or heavy, weighing 68–90kg (150–200lb). If the pigs are kept on after this they are normally for bacon, when they will have an average live weight of 72–90kg (160–200lb).

Pedigree breeding: For this enterprise it is necessary to concentrate on the more traditional breeds of pig. Each recognized breed has its own society that registers specific animals from named parents to their owners. It is an area in which small breeders have excelled, as well as saving some of the old breeds that were perilously close to extinction. Some, such as the hardy Lincolnshire Curly Coat, disappeared before they could be saved, although a similar breed still survives in eastern Europe.

Pedigree animals are more expensive to buy than hybrids, but by the same token they command a higher price when sold. The breed societies produce annual lists of breeders, and there are also specialist sales.

Raising and selling crossbreeds: Those with registered pedigree breeds will often cross a pure breed either with another breed or with a hybrid strain of commercial pig. This produces piglets that are faster growing and therefore have commercial value when sold.

Pet pigs: This is an activity that has grown in recent years. The miniature breeds of pig, such as the Vietnamese Potbellied and the Kunekune, both have their admirers. They are often kept in farm parks, where they are popular with the public.

Breeds

The decision as to which pig breed to choose is a personal one. The modern hybrids based on the Large White, Landrace and Duroc breeds are long and lean, so that the proportion of fat to meat is less than in older types. Hybrids are not as hardy as the older breeds, and may not adapt as well to outdoor conditions. Listed below are some of the breeds that are more popular with small farmers.

Berkshire: A lightweight breed with prick ears, originally developed as a pork pig. It is black in colour, with white feet and a white blaze on the face. It is hardy and also less affected by sunburn than white breeds. When the boar or the sow are crossed with a white breed, the progeny are always white.

Duroc: In terms of numbers, this is one of the most popular commercial breeds in the world. It is a hardy, ginger pig, long and lean, and an excellent sire when crossed with Large White or Landrace sows.

Gloucester Old Spots: Known as the 'cottager's pig', this breed was traditionally kept in orchards where the windfall apples provided a proportion of its food. It is a big, hardy breed with coarse hair, suitable for pork and bacon, but it is slow growing. It is popular with smallholders.

Hampshire: Introduced from America, the Hampshire is a hardy black breed with a white stripe across its back, extending to the front legs. In Britain, the breed is now referred to as the British Hampshire. It is renowned for its strong legs and leanness of meat. It is frequently used for crossing with other breeds.

Landrace: A long, lean breed originally imported from Denmark in order to improve the conformation and lean meat qualities of existing breeds. It is frequently crossed with other breeds such as the Large White, but it is not as hardy as the Duroc.

Large Black: A big, lop-eared pig, the Large Black is a docile breed that is easy to confine, for the lop

ears do restrict its vision to some degree. It has good mothering qualities, but is slow to grow. When crossed with a Large White or Landrace boar, the piglets are fast growing.

Large White: This is one of the commercial breeds frequently used for cross-breeding; outside Britain they are often referred to as Yorkshire pigs. They are suitable for pork and bacon, and are renowned for their high fertility.

Middle White: Of unmistakable appearance, with its dished face and snub nose. It is good for early pork production, and grows relatively quickly, but puts on too much fat if kept to bacon weight.

New Zealand Kunekune: The breed associated with the Maoris, and the name 'Kunekune' actually means 'fat and round' – an apt description. It is a pretty little pig, characteristically hairy, and available in a variety of colours; it is popular as a pet pig or in farm parks. It is also a useful small pig for the smallholder. It grows fast, often being ready for pork at only five months.

Oxford Sandy and Black: Related to the old Berkshire and Tamworth, this breed has only had a breed society since 1985, although its history is much older. It is regarded as a dual-purpose pork and bacon breed.

Saddleback: Known as the British Saddleback in Britain, this is similar in markings to the Hampshire, with a white stripe across its shoulders, extending down the front legs. Unlike the Hampshire, it has lop

Gloucester Old Spots sow and piglets, the traditional orchard breed and popular with smallholders.

A British Saddleback sow.

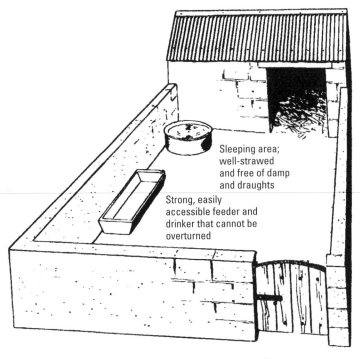

Sleeping area;
well-strawed
and free of damp
and draughts

Ridged concrete, non-slip run. The
pigs will designate one area for
dunging, making it easy to remove
the dung regularly

Strong, easily
accessible feeder and
drinker that cannot be
overturned

Strong gate

A traditional sty and run.

ears. It was developed from the old Essex and Wessex Saddlebacks of southern England. The British Saddleback is hardy, and good for cross-breeding.

Tamworth: A probable descendant of the Old English forest pig, the Tamworth has a distinctive red-gold colour and is extremely hardy. It has a long snout and prick ears, and can be difficult to confine. (It uses its nose to good effect when it comes to rooting.) It does well on rough ground, and is at home in the wooded areas of its ancestors. The Tamworth is frequently used for crossing, especially with wild boar.

Vietnamese Potbellied: This breed is similar to the Kunekune in that it breeds early and grows quickly, but it has a tendency to produce a lot of fat once it is past early pork weight. It is black and very hardy.

Welsh: Some claim that the original Welsh breed is the oldest in Britain, which is perhaps appropriate to the Welsh people who are the oldest inhabitants. It is

A movable house for weaner pigs out on range. Note the vertical flaps that allow access and exit, but help to keep the inside of the house warm. Photo: Sturdy Sties Ltd

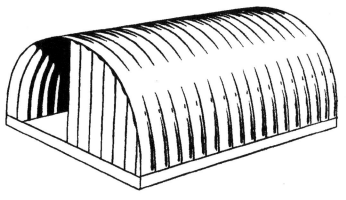

The ark may be with or without a floor but should be well supplied with straw

Movable ark for free-ranging pigs.

particularly good for cross-breeding. Its numbers became dangerously low fifty years ago, but it was redeveloped with an input of Landrace blood. For a time its breed society and that of its close relation the Long White Lop Ear of Devon and Cornwall were combined, but since the 1980s the British Lop has had its own breed society.

Housing

Housing needs to be well ventilated, yet draught free. If the roof is well insulated, and there is an ample supply of thick bedding straw, a pig will be warm enough, even in cold winters. An outbuilding or shed can be adapted for pigs, but there should always be an outside exercise yard. The concrete in a yard needs to be 'combed' to prevent slipping. Such a building can be used as a farrowing area when a sow is giving birth, or to house a boar, if one is kept. The rest of the time, free-range pigs will be on pasture. The recommended number is six breeding sows per 0.4 hectare (1 acre), with a minimum of 1.3sq m (14sq ft) inside the house. Trough space per pig should be 35cm (14in).

Rounded, galvanized metal or heavy-duty plastic arks are widely used on pasture. Many arks are designed to have a 'fender' across the door so that, while the sow can step out, her piglets are confined until they are hardy enough to go out on their own. The fender is removed when it is no longer required.

Paddock fencing does not need to be high, but strong enough to resist being pushed over or dug under; post and rail fencing with pig netting is ideal. Electric fencing is effective at controlling access to certain areas, so that pasture can be used in rotation; a battery-operated system works well, and a single strand is normally sufficient for adult pigs. Young

pigs are best confined with two strands.

Shade protection outside is essential because pigs, especially white ones, can suffer sunburn. They must also be provided with a wallow: this is simply a matter of excavating a shallow depression and keeping it filled with water in hot weather. This is the only means they have of cooling down, because they are not able to sweat, and are therefore at risk of heat stroke in high temperatures. If the wallow is made near a tree or some other shade provider, so much the better.

A trough will be needed for food, and this needs to be low for easy access, yet heavily based for stability. A stout water trough is the other necessity, because fresh, clean water must be available at all times.

We made a pig house in the corner of our kitchen

A 'fender' attached to this movable pig ark stops the piglets wandering until they are big enough to range further afield. Photo: Cotswold Pig Development Co.

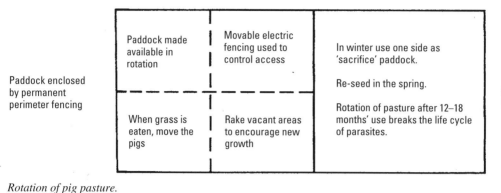

Paddock enclosed by permanent perimeter fencing

| Paddock made available in rotation | Movable electric fencing used to control access |
| When grass is eaten, move the pigs | Rake vacant areas to encourage new growth |

In winter use one side as 'sacrifice' paddock.

Re-seed in the spring.

Rotation of pasture after 12–18 months' use breaks the life cycle of parasites.

6 sows per acre (0.4 hectare)

Rotation of pig pasture.

garden out of a small, redundant brick building; it had a good concrete floor, two windows and a corrugated metal roof, and we constructed an outside enclosure of wooden stakes with pig netting. As already mentioned, pigs will do a good job of ploughing an area that is to be cultivated in the following season.

Large, intensive pig farms frequently have problems with slurry, with its associated smell, storage and pollution of water courses. This form of pollution is against the law in Britain and Europe, and people who offend in this way are subject to heavy fines. On a small scale, it is an easy matter to remove dung and urine-soaked straw regularly and consign it to the compost heap; it then goes back to the land when safely rotted down.

Feeding

The principle of feeding pigs is to ensure that they have an adequate and balanced ration for health and quick growth, but produce lean meat with a minimum of fat. For piglets and weaners, an 18 per cent protein ration is available, while growers for organic pork and bacon are usually given a 16 per cent protein mixture. Dry or lactating sows, and boars, require 15–16 per cent protein. Organic rations without in-feed medications are available, and these are the best for outdoor pigs. Compound feeds are available as pellets, cubes, rolls or meal. It is also possible to make up a ration by buying 'straights' (individual ingredients) from feed merchants, but it is difficult to get the balance of protein right for the different stages. A basic ration includes 10 parts coarse wheatmeal, 8 parts barley meal, 1 part soya meal, 1 part fishmeal, plus added minerals. The fishmeal can be replaced by soya, if preferred.

As far as quantities are concerned, the average ration per day for an adult sow and boar is 1.5–2kg (3–4lb). Lactating sows require 2kg (4lb) plus 0.25kg (0.5lb) per piglet, while piglets from 8–16 weeks need up to 1.25kg (2.5lb). From 17–21 weeks, they will require 1.75kg (4lb). Miniature breeds will need proportionately less feed.

Fodder crops can be grown for pigs, but the feeding of waste food or swill is banned. Clean, fresh water is essential at all times. During lactation and in periods of hot weather, the amount of water that a pig will need rises appreciably, and serious harm can come to them if they are deprived of water, even for comparatively short periods.

Breeding

Boars can be highly dangerous, and it is not worth keeping one for a small-scale enterprise. He needs his

Schedule 1

HOLDING MOVEMENT RECORD
The Pigs (Records, Identification and Movement) Order 1995

Name and address of person keeping the record ....

Date of Movement	Identification Mark	Number of Pigs	Premises from which Moved	Premises to which Moved

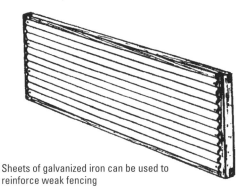

Sheets of galvanized iron can be used to reinforce weak fencing

Reinforcing weak fencing.

A wallow where pigs can cool off in summer is essential for outdoor pigs. Photo: Cotswold Pig Development Co.

own house and exercise yard, to which the sows are introduced only for servicing (though some commercial breeders run a boar outside with their breeding sows). People who specialize in the pedigree breeding of older breeds will need to have their own boar, but for most, it is much more convenient to utilize artificial insemination (AI); it also does away with the need to take a sow off the premises to an outside boar. AI service is widely available, and it is simply a matter of telephoning the breed company when the sow is on heat; the AI pack is delivered the next day, and usually contains a disposable catheter with the semen. Even the smallest breeder has a choice of top quality sires. The chances are that the sow will still be on heat, but if she shows signs of losing interest, a pheromone spray of 'boar odour' will usually rekindle her interest.

A gilt, or young female pig, is ready for breeding from about six to eight months of age. The signs of being 'on heat' are unmistakable: she may try and mount other pigs, and will stand as firm as a rock if you place your hand on her back – in fact, if you try to move her at all, you will find it virtually impossible. After opening the AI pack, all that is necessary is some non-spermicidal jelly and some paper towels, then the lubricated catheter is gently inserted into the vulva so that the semen can be injected.

Pregnancy lasts three months, three weeks and three days, give or take a few days, and during this time the sow will need to have adequate rations – although she should not be over-fed to make her too fat, as discussed in the previous section. She also needs to have adequate exercise.

A few days before farrowing, she can be treated for internal worms and external parasites, with a product such as Ivomec. This will also provide a certain level of immunity for the piglets in the first few weeks.

Farrowing

Giving birth is best carried out in the protected conditions of a building. The house should be cleaned and disinfected before occupation, and then some fresh straw provided. The sow should be introduced to her new quarters with plenty of time to settle in before farrowing. She will decide where her nest is going to be, and will arrange the straw accordingly. She will also decide which is to be her dunging area, and it is important to remove the dung every day.

It is a good idea to have a designated piglets' area with a warming lamp placed above. This is divided off from the sow's area by a metal bar to exclude the sow from the area, but still allowing her to see the piglets: this is to protect them against the possibility of being crushed by the mother when she rolls over. They go to her to suckle, but as soon as they have finished they are attracted to their little area by the warmth of the lamp.

While the sow is giving birth, it is best to be there and to place a box under a lamp, ready for the piglets. As they are born, they can be checked for deformities, and to ensure that their nasal passages are clear, given a quick rub with a towel, and placed in the box until they have all arrived. The afterbirth should be passed not long after the last piglet is born. If not, or if there is any other problem associated with the farrowing, the vet should be called: the sow may need assistance, as well as an antibiotic. (At the beginning of Part III we stressed the importance of following a course of practical training in animal husbandry and breeding before taking on the responsibility of keeping animals.)

Once the birth is over, the piglets can be gently placed near the teats, where they will immediately begin to investigate and suck. The first milk is

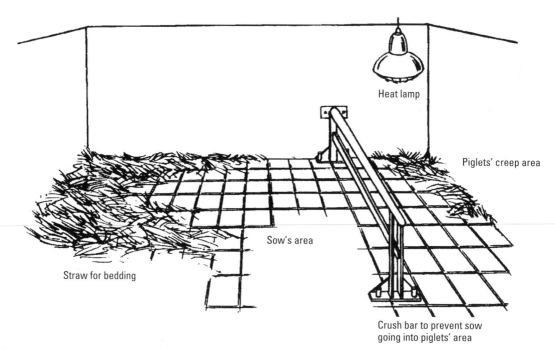

Farrowing area set up in an outbuilding.

known as colostrum and is vitally important because it is a specially concentrated food with a high level of nutrients and antibodies from the mother, which gives the piglets protection against infections and disease until they can produce their own. Occasionally, more piglets are born than there are teats to feed them, and in this case it may be possible to introduce them to another newly farrowed sow. Failing this, it will be necessary to hand-rear them, like orphan lambs. They will also learn to drink from

Heavy-based and easily accessible feeders are essential for pigs. These are some of the author's Vietnamese pot-bellied pigs.

a dish. Purpose-made replacement feeds are available from agricultural suppliers. It may be that while the sow is busy farrowing, a piglet born earlier is showing signs of distress. It is possible to give it some proprietary protein, mineral and vitamin solution that can be squirted directly into its mouth, to give it an energy boost.

Outdoor piglets do not require iron injections, nor is it necessary to castrate them if they are being sold as pork pigs. However, all pigs need to be marked for identification with ear tags, or tattoos on the inside of the ear. Breed societies also have their own systems of ear notches, depending on the breed.

Rearing, Weaning and Finishing

For the first two days after farrowing, the sow should not be given more than her normal ration of concentrates, but from the third day onwards she will need approximately 2kg (4lb) of rations a day, plus an extra 0.25kg (0.5lb) for every piglet she is suckling. Keeping the initial few days feeding as normal prevents an onrush of milk that may deprive her body of calcium. This is a condition known as hypocalcaemia, or milk fever, and in its event an immediate injection of calcium is essential. After this, recovery is equally immediate.

The piglets will be content with milk from the

A Tamworth sow with new-born piglets. Note the lamp and crush bars in the piglets' area where they will go to keep warm when they have finished feeding.

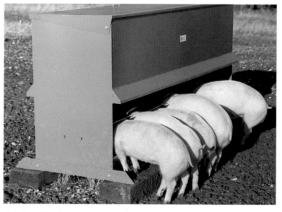

Piglets feeding from an outdoor feed hopper.

mother for two weeks, but by the third week, they will be ready for a beginner's ration. As soon as possible, but depending on weather conditions, the sow and piglets can go out on pasture. Clean fresh water should be available to them at all times. It may be necessary to provide a small drinker for the piglets.

It is important to watch out for signs of scouring (diarrhoea) in the piglets. Anti-scouring preparations are available from suppliers, but if the condition does not improve rapidly, veterinary advice is essential. The sow should also be checked for any indication of hot, swollen teats, for it could be mastitis, a condition that requires an antibiotic from the vet.

Normal weaning time is around eight weeks. By this time, the sow's milk supply will have diminished considerably as the piglets eat more solids. The sow and piglets should be moved together to the area that the piglets are to occupy. Then, after a couple of days, remove the sow and confine her in an area where they cannot hear each other. The weaners will need to be fed twice a day. A dry compound ration, with water nearby, gives all the weaners the same chance at the troughs, and is less likely to cause scours than a wet mash. The traditional advice is to give them enough food that will be cleared in twenty minutes. From weaning time to finishing age, when pigs will go to market or the abattoir, they will have been taking from 1.25–1.75kg (2.5–4lb) per pig, per day, depending on the age. Miniature breeds need less. Pure breed pigs can often be sold through the breed society.

Moving Pigs

There are occasions when a pig has to be moved from one area to another, or into a trailer for transporting to a show or to an abattoir. Forward planning is essential, with several people armed with pig boards – plain sections of board with a handhold. A pig will tend to dive through a space, even between legs, but a wide barrier such as a board can be used to control the direction it takes. The best incentive to make a pig move is to have a bucket with some food in it, so that hopefully it can be led, rather than driven. Where show pigs are concerned, training is essential if they are to show themselves to best advantage in the show ring.

To persuade a pig to go up the ramp into a trailer, it is much easier to use hurdles and make a 'race' or fenced-off track: as the pig advances, the hurdles can be closed behind it so that it is only possible for it to go forwards. Again, a bucket of pellets is a great incentive, and having several helpers does make the task much easier.

In Britain, a movement licence must accompany a pig when it is moved from one site to another. Blank copies for the owner to fill in are available from the local Animal Health Office (*see* page 178).

Health

Avoiding trouble is largely a matter of common sense. It has already been pointed out that pigs are essentially clean animals that rely on humans for proper care and management. Outdoor pigs are less likely to suffer from the range of respiratory and viral conditions that can affect intensively reared ones. Any bought-in pigs should be kept apart from other livestock for at least ten days, in case they are incubating an infectious disease.

It is also a good idea to isolate an animal that is ill, providing a quiet, well strawed area for it to rest, as

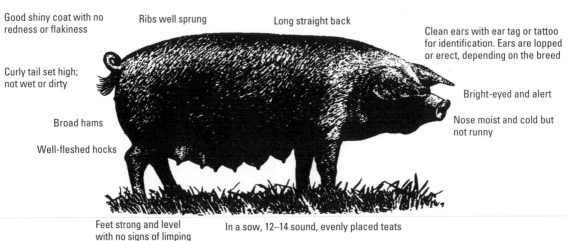

Good shiny coat with no redness or flakiness

Ribs well sprung

Long straight back

Clean ears with ear tag or tattoo for identification. Ears are lopped or erect, depending on the breed

Curly tail set high; not wet or dirty

Bright-eyed and alert

Nose moist and cold but not runny

Broad hams

Well-fleshed hocks

Feet strong and level with no signs of limping

In a sow, 12–14 sound, evenly placed teats

What to look for in a healthy pig.

well as a drinker and feed container with a few concentrates. Veterinary advice should be sought in all cases of illness that show no signs of clearing up, or where there are unexplained symptoms: these include shivering, panting or irregular breathing, poor appetite, constipation, sneezing, vomiting, diarrhoea, discoloration of the skin, swollen navel, udder or joints, and lameness.

Notifiable Diseases
Foot and mouth disease, and swine fever, are both highly infectious viral diseases. The former can affect all cloven-hoofed animals. Symptoms include loss of appetite, high temperature, lameness and some slavering at the mouth, although this is more noticeable in cattle than in pigs. Animals do recover from the disease and it is possible to vaccinate against it, but official policy is still to slaughter affected herds and compensate the owner.

Swine fever is also caused by a virus and tends to affect young pigs. Symptoms include shivering, loss of appetite, excessive thirst and possibly vomiting. There may also be a purple rash on the ears. The death rate is high. The presence of both these diseases must be notified to the authorities.

Vaccinations
An important vaccination is against erysipelas, a disease that can manifest in several forms, from mild to acute, and can lead to permanent heart trouble and damage to the joints. Other vaccinations are available for a range of respiratory diseases, but these are more applicable to intensively housed pigs. Free-

range animals kept on a small scale are far less likely to be affected.

It is necessary to keep a record of all medicines that are administered.

Foot Care
Breeding stock will need to have their claws clipped from time to time, and as they are strong, professional aid may be needed, unless the pig-keeper is skilled and experienced in the technique. Purpose-made clippers and rasps are available from suppliers.

Worming
Worming should take place after weaning, and at regular intervals afterwards. A product such as Ivermectin will deal with all types of parasitic worm, but the vet will advise on suitable preparations, as well as on vaccinations.

External Parasites
The skin of a pig is liable to a number of conditions, particularly lice and mite infestation. Ear canker from mites can also be a problem. Patches on the skin should be watched for. In all cases, cleaning the affected area is a matter of priority, followed by the application of a proprietary medication.

Mastitis
This is an inflammation of the teat after farrowing. A careful watch should be kept for symptoms such as a hot, red udder that is painful to the touch. The vet will prescribe penicillin to clear it up.

15 Sheep

Feed my lambs, feed my sheep. (Gospels)

Sheep are kept for their wool and fleeces, for their meat or, in some specialized cases, for their milk. In Britain and the USA it is necessary to register with the authorities if sheep are kept, as well as conforming to animal welfare, handling and identification legislation.

The easiest form of sheep keeping is where there is no breeding involved. Here, a mixed flock of ewes that are not put to the ram can cohabit peacefully with wethers (castrated males), and kept purely for their wool. This is often the practice of those whose main interest lies in wool crafts such as spinning, weaving and dyeing. A small number may be raised for meat, or kept purely for keeping down the grass in an otherwise unused paddock.

A breeding flock requires more time and management. A good ram is essential, although rams can often be hired. It is important to remember that they are often aggressive, with horned breeds being capable of inflicting considerable damage. Small-scale enterprises are best advised to keep the traditional breeds, for there is a growing interest in their conservation, with good quality breeding stock being sought after. Improving a breeding flock by judicious selection will produce a good strain of a particular breed. It is also common practice to cross a pure-bred ewe with a commercial meat ram so that the progeny can be sold as meat lambs.

Breeds

Different breeds of sheep have been developed for different terrains. In hilly areas, the sheep are generally smaller and hardier than the lowland breeds. It is a good idea to select those that are suited to a particular environment. Also important is to choose a breed that is appropriate for the handler: a large, heavy breed may not be the best choice for a small person who is not physically strong. It is impossible to list every breed available, but sheep are classified as follows.

Teeswater, a large, hornless breed of longwool sheep.

Prize-winning Southdown ram lambs at the Surrey County Show. Note their halters and leads.

Longwool breeds: These are big sheep with heavy fleeces and long, frequently lustrous wool that formed the basis of the worsted trade. Examples include Cotswold, Romney, Leicester, Lincoln, Teeswater and Wensleydale.

Shortwool and Down breeds: The traditional shortwool breeds of the Middle Ages have given rise to the Down breeds of today. They have shorter, dense fleeces for wool production. Examples are Suffolk, Dorset Horn, Hampshire Down, Southdown, Dorset Down, Shropshire and Ryeland.

Hill breeds: The traditional mountain sheep are very hardy. There are two main groups: the *white-faced*, such as the Welsh Mountain and the Cheviot, and the *black-faced*, such as the Scottish Blackface and the Swaledale. They are used to ranging over a wide area, and can be difficult to confine.

There are also sheep that have originated from mountain breeds, but were crossed with other breeds over the years; examples are the Lleyn and the Clun Forest.

Rare breeds: These include the old, primitive breeds that are generally extremely hardy and do well on poor, sparse pasture. They include the Soay, Hebridean, Manx Loghtan, Boreray, North Ronaldsay, Castelmilk Moorit and the Shetland. There are other breeds that are of more recent development, but are categorized as rare because they may no longer be regarded as commercial breeds. They include some hill and down breeds.

Coloured wool breeds: Most of the coloured wool sheep are found amongst the primitive breeds, or the black breeds of Wales that include Black Welsh Mountain, Balwen and Torddu. The Herdwick and Gotland have grey fleeces, while Icelandic, Shetland and North Ronaldsay are available in a wide colour range. The Jacob is pied, with black spots on a white fleece. They are all popular with those interested in wool crafts.

Normal white breeds can also throw up recessive-gened black lambs. These have no commercial value as far as large sheep farmers are concerned, but smallholders have used this factor to breed coloured varieties of breeds such the Ryeland, Merino, Leicester and Corriedale.

Dairy breeds: The Wensleydale and Dorset Horn were the traditional milk producers, though these days, the Dutch Friesland and British Milksheep are the ones used commercially. The Lleyn, Llanwenog

Badgerface Welsh Mountain sheep of the Torddu type.

Clun Forest sheep originated from mountain breeds, but were crossed with other breeds until they were standardized. Note the identifying ear tags.

The Jacob is an ancient breed, popular with spinners and weavers for its fleece that produces naturally coloured wool.

and Texel are also suitable for milk production.

Meat breeds: The two main meat breeds are the Dutch Texel and the French Charollais. These are often used to cross with pure-bred or cross-bred ewes, with the lambs going for slaughter.

Sheep Handling and Housing

For a large flock, a shepherd will probably use a dog to help with rounding up, particularly where sheep are grazing in unfenced mountain areas. Smaller flocks can normally be handled without such assistance, but enclosures, hurdles and pens are essential. Food is also a great incentive, and small flocks frequently come running when they see or hear their owner.

Once sheep are penned, they are accessible for whatever purpose, but they may also need to be caught and cast. This involves grasping the wool under the throat with one hand, while the other hand gets hold of the wool just to one side of the tail. Lift the back of the sheep, while simultaneously thrusting with your knee, and it will lose its balance, enabling it to be placed in a sitting position leaning slightly to one side. In this position its feet can be examined, and it is also ready for shearing. Heavily pregnant ewes should not be cast in this way.

Sheep are essentially creatures of the outdoors, but having some kind of shelter for lambing makes life easier for the owner, as well as saving the lives of a good many lambs. More delicate breeds, such as milk sheep, need shelter for most of the year. Well ventilated, open-fronted barns or sheds are ideal, while reference has already been made to the increasingly common use of shaded polytunnels. Allow at least 1.5sq m (16sq ft) per ewe, with 50cm (20in) trough space and 15cm (6in) hayrack space.

Sheep hurdles are virtually indispensable to the sheep enterprise. They can be used to make temporary pens for all manner of reasons, and are particularly useful in the construction of individual pens within the building where sheep are brought in to lamb. They are invaluable in making handling pens, gates and 'races' for controlling and directing sheep to particular areas.

Fencing is important, as some breeds – notably the mountain sheep, and some of the primitive breeds such as Jacob and Soay – are notorious for their ability to escape. Ideally there should be permanent fencing around the site as a whole, with moveable electric fencing for controlled access to grazing. Electric netting should not be used for horned sheep because of the danger of entanglement.

A polytunnel in use as a sheep shelter. Note the ventilation panels at the base, and the metal hurdles used to make pens.

Feeding

Sheep are grazing and cudding animals. The latter process is when a certain proportion of relatively quickly grazed grass is regurgitated and chewed at leisure, usually when the sheep is at rest. The most important part of the sheep's diet is grass, and they will spend most of the year grazing. The nutritional value of grass is at its highest in the spring and early summer; when the grass declines, they will need supplementary feeding. Hay is the main feed source, but pelleted feeds are also used at specific periods such as at tupping, in winter, and when lambing. Sugar beet pellets are useful for providing energy, and can be scattered on the grass. Cereal-based pellets or cubes provide the extra protein, minerals and vitamins for lambing and lactating ewes.

Concentrate feeds normally have a protein content of 14 per cent, 16 per cent or 18 per cent. The first is suitable for general use, while the latter two are more appropriate for lambing. Organic feeds are also available. Troughs for the feeding of concentrates should be low enough for access without allowing food to be contaminated by droppings. Hayracks are useful in the winter housing or out on the field.

Milk sheep will need a concentrate ration all through the year, in addition to grass and hay. This will be somewhere between 450g–1.3kg (1–3lb) a day, depending upon the level of milk production.

Breeding

The choice of ram is important, for he will have a 50 per cent determination factor in the quality of lambs the following season. His feet will need to be in good condition, otherwise he will not be able to service all the ewes. It is important to check with the vendor that the ram is being sold as 'warranted fertile'. If you buy on this basis, then you can return him and get your money back if he fails to get the ewes in lamb. Check his teeth to see how old he is, as well as his general condition. Both testicles should be descended.

Where a ram is already established on the farm, he should be checked over in good time before meeting the ewes. This allows any problems with his feet, or anything else, to be cleared up. Rams are more docile out of season (although children should always be kept away from them), but from autumn onwards it is best to pen them on their own, or with an old ewe for company. As they come into season they become aggressive. While they are penned, they will need hay and green food, but it is important not to overfeed them as that will make them fat and lazy.

Ewes that are selected for breeding should be sound and healthy, and have a good udder. Late summer is a good time to purchase breeding stock, and it is important to buy ewes which come from a flock with a record of good lambing and rapid weight gain. It is also important to assess the body condition and check the feet of the sheep.

Sheep and chickens frequently share a pasture to their mutual benefit.

Flushing

Flushing is a system of feeding up the ewes before they meet the ram. It applies to newly bought ewes, as well as ewes from an existing flock that have been grazing all summer. They are given richer pasture to bring them up to peak condition so that multiple ovulation is more likely, resulting in more lambs. A good way of providing pasture for flushing is to turn the ewes out onto grass that was cut for hay earlier in the year, and which has been allowed to grow on as 'foggage', or late grazing. Sugar beet nuts can be scattered on the grass, or a feed block made available. A concentrate ration of 250–500g (9–18oz) per ewe is appropriate at this time.

Just before the ewes are turned out onto their new pasture they should be wormed. As they are being handled, it is a good opportunity to make another quick check on their feet and on their general condition, because once they are pregnant, handling should be kept to a minimum.

Tupping

This describes the period when the ewes run with, and are covered by, the ram. The rams can be fitted with a sire harness strapped to the chest; this carries a raddle block, a colouring agent, so that as he mounts the ewe during mating, a mark is left on her rump. It is a method of checking which ewes have been served and which have not. Different colours are used, with the colour being changed every seventeen days so that the date of lambing can be worked out. If ewes that have already been served are seen to have a second colour, this is a sign that they have come into heat again and were not pregnant the first time. If this is widespread, it is a sign that the ram, although active, is not very fertile. Where a flock is made up of different breeds there may be a tendency for the ram to select some and not others. During the period of tupping, a careful eye needs to be kept on what is happening, to make sure that all the ewes become pregnant. A young ram can serve up to about twenty ewes, while an older one can manage up to forty. The normal tupping period is about seven weeks. Concentrate feeding should be 250–500g (9–18oz) per ewe, in addition to its grazing activities.

Before tupping, the areas of wool around the ewe's tail can be clipped back a little in case it impedes the ram; this is known as 'crutching'.

The Pregnant Ewe

Pregnancy lasts for approximately five months, and during this time the ewes should be fed adequately to cater for their needs, but without over-feeding and risking making them too fat. Keep to the same diet

and feeding level as when tupping took place.

Once the weather worsens, it is a good idea to bring the ewes to a sheltered area. Hay, fed ad lib, will be needed to replace the grass. Concentrates should be given during the last two months of pregnancy, 'steaming up' the ewes to ensure that the growing lambs are provided for, starting at 250g (9oz) per day and increasing steadily to 900g (34oz) per day during the last two weeks. This should be divided into two feeds a day.

About 4–6 weeks before lambing, the ewes should be vaccinated against clostridial diseases and pasteurellosis; this will also confer immunity to the lambs for around ten weeks. From 10 weeks, the lambs should be vaccinated, with a booster injection following 4–6 weeks later. Thereafter, vaccinations should be carried out yearly, not forgetting the ram.

Lambing

In addition to the area prepared for ewes about to lamb, a 'sick pen' should also be made ready. This is a pen with a heat lamp and large cardboard box underneath it so that any weak, hypothermic or orphan lambs can be revived.

A ewe that is about to lamb is restless and will move away from the rest of the flock. She frequently paws the ground. Where the facilities exist, it is better to separate her from the others by putting her in a pen on her own. While the sheep is giving birth, she adopts a characteristic position with her head pointing straight up at the sky. Unless there are obvious difficulties, there should be as little interference with the birth as possible. Occasionally a breech presentation occurs, or the lamb may be a particularly large one. In situations of this kind, it is necessary to obtain the help of an experienced sheep

A new-born lamb with its mother. There is less chance of disease with outdoor lambing, but a higher risk of hypothermia.

New-born lambs under a heat lamp.

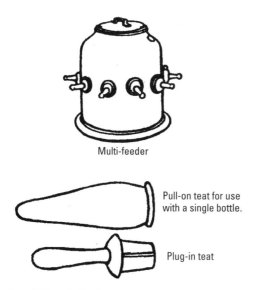

Multi-feeder

Pull-on teat for use
with a single bottle.

Plug-in teat

Lamb or kid bottle feeding.

handler or vet. Again, it is worth reiterating how essential it is to have attended a practical husbandry course that includes lambing. This will include information on what to do in an emergency, when to call the vet, and what equipment and medications should be assembled beforehand. If, for example, it is necessary to assist the ewe, wear disposable gloves and use plenty of lubrication.

Once the lamb is born, the mother usually licks it vigorously, after which it stands and starts to suckle. Check that it is receiving milk by feeling the udder to ensure that it is emptying. If it hard, this is an indication that it is not emptying, a situation that could lead to mastitis. If the lamb is feeding, its stomach will feel warm and distended. The first milk, or colostrum, is vital as it contains large amounts of nutrients and antibodies that help to protect the lamb against disease. Treating the navel of a new-born lamb with tincture of iodine BP also protects it against infection.

The ewe's afterbirth should be removed, and fresh bedding straw supplied. The ewe should be given

Bottle-feeding an orphan lamb.

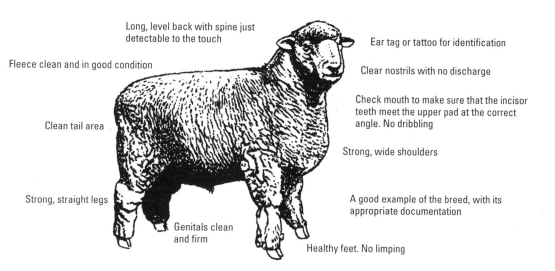

Long, level back with spine just detectable to the touch

Ear tag or tattoo for identification

Fleece clean and in good condition

Clear nostrils with no discharge

Check mouth to make sure that the incisor teeth meet the upper pad at the correct angle. No dribbling

Clean tail area

Strong, wide shoulders

A good example of the breed, with its appropriate documentation

Strong, straight legs

Genitals clean and firm

Healthy feet. No limping

What to look for in a healthy sheep.

hay and plenty of fresh water to drink. The first twenty-four hours are vital for the bonding process between ewe and lamb. Keep a watch for any signs of milk fever or mastitis, as referred to earlier.

Sometimes one is faced with the question of what to do with orphan lambs. If another ewe has just lambed and lost one of her own, it may be possible to introduce an orphan to her, but this must be done as quickly as possible. The orphaned lamb must be rubbed with some of the afterbirth of the foster mother. This may persuade her to accept it, thinking that it smells of her own. However, lambs are very difficult to foster with other ewes, and hand-rearing is often more successful. It should be given some colostrum from one of the other newly lambed ewes, to have a reasonable chance of sturdy growth. Bottle-reared lambs are undoubtedly attractive little creatures, but the problem is that when they get

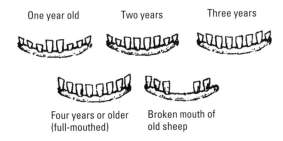

One year old   Two years   Three years

Four years or older   Broken mouth of
(full-mouthed)   old sheep

Establishing the approximate age by examining the teeth.

older, they never really adjust to the flock, and follow their owner everywhere. Automatic lamb feeders are available that work on the principle of a central tank filled with milk substitute, the lambs sucking attached teats. It is not too difficult to construct such a contraption oneself, and it can also be used for feeding goat kids. Milk substitute is available from feed suppliers, and the instructions should be followed carefully.

The sheep is particularly vulnerable to blowfly attack. The tail tends to become soiled with dung, and in the summer this can become a breeding ground for maggots. It is normal practice, therefore, to dock the lamb's tail after the bonding period of the first day. It must be done in the first week of life. The way to do this is to use an elastrator to apply a rubber ring around the tail; as it tightens, the blood supply is cut off and in about a week the tail drops off. Male lambs are castrated by applying another ring around the scrotum; however, this is not necessary if they are to go for slaughter at around four months of age.

Identification of sheep is necessary. It can be by ear tags, colour markings or both. Small tags are available for lambs, but need to be replaced by normal ones later.

Rearing Lambs

For the first two or three weeks the lambs will feed exclusively on the mother's milk, so the provision of adequate water, grass, hay and concentrates for the ewes is vital. If lambs are being sold early, they can be given 'creep' feed from three weeks old: small, high-protein nuts placed in a feeder that is not accessible to the adults. They will also take hay from an early stage. The ewe's concentrate ration will gradually decrease, until six weeks after the birth it is no longer necessary. After about a week, the ewes and lambs can be released from the pens into a small sheltered paddock, and eventually into open fields.

Lambs for meat can be sold at local markets or to the abattoir. Pure-breed lambs for other purposes are best advertised through the breed society or small farming magazines.

Sheep shearing normally takes place in early summer, and coincides with the growth of new wool after the winter (*see* page 216).

Dairy Sheep

Dairy sheep can be hand- or machine-milked without any problems, although for hand-milking they will need a platform to raise the level to a comfortable height for the milker. (For details of hand milking *see* page 201, while further information on machine milking is to be found on page 209.)

Ewe's milk has a high butterfat content, similar to the milk of Jersey cows, but the smaller fat globules make it easier to digest. Like goat's milk, it can be frozen. A special feature of ewe's milk is its high casein content of 5 per cent, against 2.5 per cent and 3.2 per cent in goat's and cow's milk, respectively; this means that 1kg of cheese can be made from 4 litres of milk, as compared with 9 or 10 litres from goats or cows. It is also good for yoghurt and ice-cream production. Any commercial dairying operation will require the premises to be registered and inspected for compliance with the dairying, food safety and packaging regulations.

Health

When acquiring sheep for the first time, it is a good idea to buy those from a flock that has been monitored for scrapie disease: this will ensure against the possibility of having a BSE-type problem with sheep in the future.

Condition Scores	
Score	Comment
0	No muscle or fat between skin and bone. Sheep emaciated and close to death.
1	Vertebrae prominent and sharp. Muscle thin with no fat cover. Very lean.
2	Vertebrae still prominent. Muscle is of moderate depth but little fat cover. Lean.
3	Vertebrae feel smooth and rounded. Muscle full with moderate fat cover. Good condition.
4	Vertebrae only detectable as a line. Muscle full with thick covering of fat. Fat.
5	Not possible to detect vertebrae. Grossly fat.

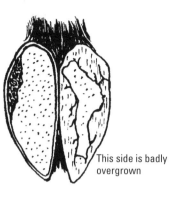

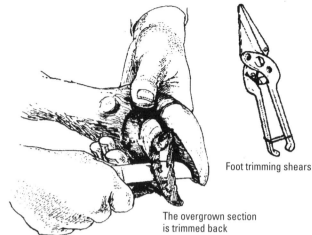

Foot rot indicted by stinking pus between outer horn and soft inner area

This side has been trimmed

This side is badly overgrown

The overgrown section is trimmed back

Foot trimming shears

Foot care for sheep and goats.

Sheep need to be checked regularly for signs of illness, fly strike or foot problems, in addition to their normal care. It is good practice to visually check for unusual signs every day, as well as carrying out the periodic checks listed below.

Vaccinations

Sheep can be vaccinated at 4–6 weeks before lambing, against the range of clostridial diseases that include enterotoxaemia, pulpy kidney and tetanus, as well as pasteurellosis. The lambs should be vaccinated at 10–12 weeks old and given a booster vaccination 4–6 weeks later.

Worming

Ideally, rotation of pasture should control worms, but if the ground is overstocked, regular worming is required, with the frequency depending on the quality of the grass and its rotation. If sheep are scouring it may indicate a worm burden. The ewes and ram should be dosed at tupping time when they are introduced. The sheep will need to be gathered into a race of pens so that movement is restricted. Place one hand under the sheep's chin and insert the nozzle of the dosing gun in the back corner of the mouth. Press the trigger steadily so that the animal can swallow easily.

Ewes can be wormed again before they are housed for the winter, and then three weeks after lambing. Lambs should be wormed at six weeks, and then at regular intervals. It is best to get veterinary advice on this, for local conditions will vary. The vet will also advise on worming products, and the need to change them in case immunity is built up by using the same brand.

Foot care and regular hoof trimming is an important part of livestock husbandry.

Teeth

An examination of the teeth on the lower jaw will reveal the age of the sheep. (The upper jaw has a bony pad). A one-year-old will have two incisors at the front; these will be seen to be bigger than the remaining teeth. Such an animal will be a first-time breeder or yearling ewe, and is the best to purchase for a future breeding flock. As sheep get old and lose their teeth, they are referred to as 'broken-mouthed', and it is difficult for them to graze effectively.

Foot Care

The feet need regular trimming, for the hoof cleats can grow round and over the soles of the feet if not checked. Foot rot may develop if bacterial infection affects the sole. Affected sheep should be separated from the flock for twelve days as the bacterium cannot survive for longer than this outside the hoof. Once the infected area is cleaned out, the sheep should stand in a foot-bath solution containing zinc sulphate or copper sulphate for several minutes.

Condition Scoring

This is a method of assessing body condition by checking the area around the spine, behind the last rib. Using the fingertips, it is possible to determine the amount of muscle and fat covering the vertebrae, and then giving each ewe a score on a scale of 0–5, as indicated on page 190. Around 3 is the ideal (although mountain and primitive breeds tend to be thinner). Ultra-thin or obese animals should have their feeding adjusted accordingly.

Fly Strike and Sheep Scab

Sheep can be protected against fly strike and other insects, lice, mites and ticks by dipping or spraying after shearing. Lambs can be sprayed before going out on pasture. These days, there are alternatives to the organo-phosphorus compounds that were used before their danger to human health and to the environment were established. Flumethrin is a synthetic pyrethroid that can be used, but care must be taken with its disposal so that it does not go into watercourses, for it will kill fish. An organic recommendation is to neutralize it with slaked lime after use, and then stir it every few days; after a fortnight it can be diluted with manure or slurry and spread on grassland, away from ditches, streams and other watercourses.

16 Goats

Goats have slit eyes to enable them to see round corners.

The goat is a ruminant browser rather than a grazer. In the wild, its ability to eat large amounts of vegetation and then chew it later, at leisure, is a survival technique to protect it from predators. Its preference for shrubby growth and broad-leaved weeds, rather than grasses, provided a wide range of nutrients and minerals from its ranging activities on the hillsides. Eating higher-growing plants also protected it from many of the parasites and infections that are more prevalent at ground level. The fact that goats will only eat hay from a rack and will ignore that on the ground is a reminder of its wild origins.

Goats are kept as dairy animals; to a limited degree for meat; as fibre-producing animals; and as pets. They are social animals, and need the company of their own kind. They are hardy, but have a dislike of the rain so must have sound, well ventilated shelter. When buying goats it is advisable to acquire animals that come from registered, pedigree stock and so have their own registration card with all its details, including the signature of the last owner. Without this, it will not be possible to register the transfer of ownership. If they are dairy animals, they should ideally be from milk-recorded stock, and they should certainly be from a CAE-tested herd: this means that all have been tested for caprine arthritis encephalitis, a viral condition for which there is no cure. Tested and monitored herds can qualify for CAE-free status.

Anyone with goats must register with the local Animal Health Office, and meet the appropriate animal welfare and management legislation. The goats need to be identified with ear tags or tattoos.

The British Goat Society is the body that co-ordinates goat-keeping activities in Britain and will put people in touch with goat-breeders in their particular locality. In the USA, it is the American Dairy Goat Association. Local goat societies that are affiliated to the national organizations are also excellent sources of help and advice.

Breeds

There are many breeds of goat in the world, some feral, and some developed for specific production with differing names in their own localities. The following are the main breeds that are widely available.

Saanen: The all-white Saanen originated in Switzerland from where it has been exported all over the world. It is a smaller animal than the British Saanen and the American Saanen that were both developed and standardized from it by upgrading with local strains. These produce the highest levels of milk.

Toggenburg: Also from Switzerland, the original Toggenburg is a relatively small goat, brown in colour with white stripes on the face. The coat is often soft and silky, with a proportion of long hair. The British Toggenburg developed from it is much larger and has better milk production. In the USA and elsewhere the Toggenburg has also been cross-bred and upgraded for higher milk yields.

Alpine: Alpine goats were introduced to Britain in 1903 and were used to grade up local goats. The

Florence, the author's British Saanen goat, in her pen. Photo: Chris Sowe

Pygmy goats are frequently kept as show or pet animals.

British Alpine goat demonstrating that a goat always wants to see what's on the other side of the fence.

Boer goat. This is the best breed for rearing as meat. Photo: Boer Goat Society

result is the British Alpine, which is glossy black with white face stripes, and in some cases a white belly. There are French and Italian Alpines of varying hues on the continent of Europe. The French Alpine is also widespread in the USA.

Nubian: This distinctive breed with its Roman nose and long, droopy ears was bred from indigenous British goats crossed with Indian and Sudanese Nubian goats. In Britain it is called the Anglo-Nubian. In the USA and other countries to which it has been exported, the prefix 'Anglo' has been dropped and it is simply called Nubian. Its milk has the highest butterfat content, and the breed has been called 'the Jersey of the goat world'.

Pygmy: These are small African goats that are popular as pets. They have active supporting societies in Britain and the USA, and many are exhibited at country shows. They are not disbudded, as is the practice with dairy goats, and have their small horns left intact. They respond well to training and will walk on a lead.

Boer: The Boer goat was developed as a meat goat in South Africa. It is a short-legged, stocky animal with a broad chest and rounded body. It is hardy and has no problem with damper climates. Boer goats are docile and can be managed like dairy goats. Females can feed their kids at foot, but do not produce a surplus for milking. A kid is normally 4kg (9lb) at birth, and will reach 30–40kg (66–88lb) at six months old. They can be killed out at this age, producing a 50 per cent carcase weight of 17–21kg (37–46lb). The numbers of imported animals is still low, and there has been some controlled cross-breeding to build up the numbers. The British Boer Goat Society brings breeders together, maintains a herd book and registers the goats.

In the USA, range-reared meat goats are usually referred to as Spanish, although these now bear little resemblance to the small, brown and highly productive dairy goats of Spain.

Other breeds: The Golden Guernsey is localized in its distribution. It is a pretty animal with golden, silky hair that is often long and wavy. The English

Guernsey was developed from it using Saanen and British Saanen bloodlines: it is based on indigenous animals before the introduction of Swiss breeds. In the American La Mancha, the most noticeable feature is the absence or reduction of the external ears. The Bagot goats of Britain are small black and white animals with a history of at least 600 years. There are also feral populations of Cheviot and Welsh goats in the mountains. Finally, Angora goats are producers of fine quality mohair; more details of these are given in Chapter 18, for their management is more akin to that of sheep and alpacas than to that of dairy goats.

Housing and Management

If goats are kept on range, ten goats per hectare (four per acre) is an average herd density, although this depends on the quality of pasture and on efficient pasture rotation. Dairy goats need housing that is dry, well ventilated and free of draughts. Doors and gates need to have firm bolts, for goats are intelligent and are adept at opening them. There are many possibilities when it comes to housing, depending on the scale and nature of an enterprise, but the basic needs do not vary and may be listed as follows: sleeping area, milking area and ranging area.

Sleeping Area

Commercial herds often have a communal barn for their dairy goats or for those of different ages: a minimum of 1.4sq m (15sq ft) should be allowed per animal. The other option, and one that is usually preferred by those with a small number of animals, is to have a separate stall for each goat, but so designed that the goats can see each other from their individual stalls. A convenient stall is 1.8 × 1.2m (6 × 4ft), with the door on one of the short sides.

The barn or stalls will need a hayrack to hold hay and green food, and a securely supported water bucket. A convenient way of refilling an individual bucket in a stall is to have it placed in a gap in the door, at the right height for the goat. The floor will need a layer of bedding straw to absorb droppings and provide a warm, insulated surface on which to rest. This will need periodic removal and replacement, with the old litter going onto the compost heap.

Milking Area

The milking area is obviously separate on grounds of hygiene. It should be light, airy and easy to hose down. As far as goats are concerned, a milking stand is useful so that the animals are at the right height for those milking them. On a domestic scale, milking may be by hand. Where milk is being sold, machine milking is required, and food safety legislation must be met. This will also entail having registered premises that are inspected.

Ranging Area

This may be an area of pasture, an enclosure, or even an exercise yard. Access to shelter in the event of rain is required, as well as a drinker and a hayrack for hay or other browse material. If an exercise yard is used, it is a good idea to have some blocks of wood, tree trunks or flagstones for the goats to climb on, for they are descended from creatures of the heights. This is particularly popular with goat kids.

As far as our goats were concerned, we worked

A small goat house suitable for two animals. Note the window that provides ventilation but excludes draughts.

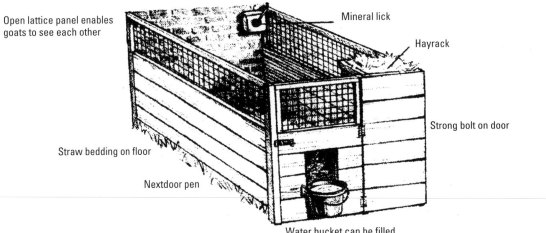

Open lattice panel enables goats to see each other

Mineral lick

Hayrack

Strong bolt on door

Straw bedding on floor

Nextdoor pen

Water bucket can be filled from the passage outside

Individual goat pen in an outbuilding.

out a system that catered for all these needs and saved us as much time as possible. We converted part of the range of outbuildings in such a way that what were originally two interconnected buildings became two pens, a passageway, feed store and a separate milking area with a sealed concrete floor and drain.

We made a wooden platform that stood 45cm (18in) off the floor. At one end of this was a feed bucket in a fixed metal stand with a short length of tethering chain attached to the wall. The goats were given half their daily concentrate ration while they were being milked, so that they stood contentedly. Outside, we made a small enclosure with a weather shelter in one corner and some flagstones for climbing. Enclosed by a post and rail and netting fence, it had a gate for access. During the day, the goats could run around unimpeded. A large metal hayrack on the outside wall of the shelter was periodically filled with browsing material such as twigs and other hedge material that would not spoil if it rained. Inside the shelter was a smaller rack for hay. A water bucket in a stand was the only other necessity.

Feeding

Goats require a balance between roughage, such as hay, twigs and other herbage, and concentrates, such as cereals that have a higher nutritional value in relation to their volume. Concentrate feeds (mixes, pellets or cubes) are widely available, including those for organically reared goats. It is useful to

think of a dairy animal as needing a maintenance ration and a production ration: the first caters for its basic nutritional needs that are provided for by hay and a 13.5 per cent protein mix. The second caters for basic needs as well as extra demands made on the system, such as milking. In this case, hay and a 16 per cent protein mix would be appropriate. Mixes should be introduced in small quantities at first, gradually building up as necessary. It is a good idea to divide the concentrate ration so that it coincides with morning and evening milking.

Our experience was that keeping the hayrack full of hay at all times gave the roughage element of the diet priority. The concentrate part, given morning and evening, was then easy to estimate by how much each animal was likely to take at each milking. As a rule of thumb, it worked well. Reference was made earlier to the fact that goats will not eat hay that has

Metal ring for inside use

Metal pronged stand for outside area

Bucket placed in tyre for outside use

Examples of bucket holders.

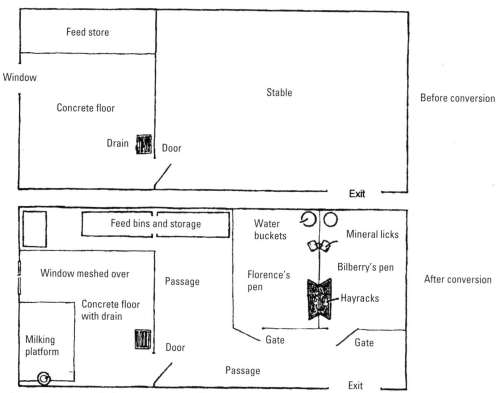

The author's goat housing made from converted stabling.

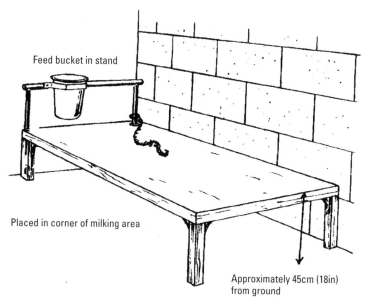

A milking platform for goats.

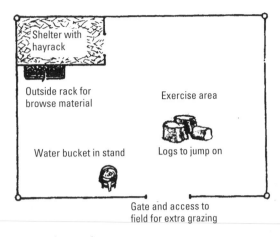

The outside corral.

fallen on the floor. The best way of avoiding wastage is to have a lid on the rack so that the hay can only be pulled out from between the bars. Haynets are not recommended for goats, for they can easily become entangled in them.

Fodder crops such as kale, comfrey, pea haulms and lucerne (alfalfa) can be given, if required. Root crops such as turnips and carrots are popular. Large ones should be chopped in case the goats choke on them. Popular browse material includes blackberry, bramble and twigs of willow, hazel, elder and apple. (Avoid evergreen branches and hedge cuttings that may include poisonous plants or berries.)

Breeding

The onset of oestrus, or the breeding cycle, takes place in autumn and the average gestation period is 150 days, although this will vary depending on the breed. The signs of being on heat are bleating and sideways tail-wagging. There are incidences of goats being 'maiden milkers' (producing some milk without having kidded), but for most purposes it will be necessary for a female to kid before a milk supply is available. A male goat is only required for those who are breeding on a commercial scale; most goat keepers will take their females to a local stud male, as required. AI services are also available.

A buck (billy) will need his own quarters, with an exercise yard and strong, high fencing. He sprays his quarters, and the smell can be very offensive. A female is introduced into his yard as necessary. If a female is taken off the premises for breeding purposes, a record of the movement must be kept, as detailed earlier. Stud fees for goats are quite reasonable, and if the goat does not 'take' the first

time, it is normal to have a free service the second time.

A pregnant goat should not be over-fed, and she should have plenty of exercise. Clean, fresh water needs to be available at all times, although it is easy to gain the impression that goats drink infrequently. When kidding time is near, she should have her own stall with fresh bedding straw. Kidding is normally straightforward. Our goats always fooled us by having their kids when we were not looking, although they were checked regularly. If there is any indication of a problem, the vet should be called.

Male kids have little commercial value, as there is little demand for goat meat, and are normally put down at birth. This must only be done by a competent person. If a vet comes out to do it, he can be asked to disbud the female kids at the same time (*see* below). All goats are required to be identified in the same way as sheep, by tagging or tattooing.

For the first few days, it is best to give hay with just a few concentrates to the mother; about 0.2kg (0.5lb) is enough. This reduces the possibility of a sudden rush of milk leading to a calcium deficiency, as referred to earlier. From the third day onwards, the concentrate ration can be increased to around 0.45kg (1lb), and then gradually increased to a maximum of 2kg (4.4lb) as her milk production increases. At all times, the balance of roughage and concentrates needs to be maintained.

With a dairy goat, the female kids should be allowed to feed off the mother for five days before

Goats are keen on branches and other browsing materials, as this British Toggenburg indicates.

Average Feed Requirements		
	Hay	*Concentrates*
Adult goats	1–2kg (2.2–4.4lb)	0.5–1kg (1.1–2.2lb)
Milking goats	3.5kg (7.7lb)	2kg (4.4lb)

Note: Young animals and small breeds will eat less. Amounts also vary depending on production levels.

being bottle-fed; at this stage the kids are put in their own pen where they have each other for company. Dried milk substitute is available. The first few days of the mother's milk is vital to the young, because it is colostrum and contains antibodies essential to the health of the kid.

At first the kids will take not much more than 0.12 litre (0.25 pint) at each feed, but this will gradually increase until they are drinking around 2 litres (4 pints) a day. Initially four feeds a day will be necessary, but by the time the maximum amount is reached, three daily feeds will be sufficient. It is never too early to introduce hay. In my experience, kids enjoy having a few tufts, even if it is only to play with in the first few weeks. Nevertheless, it accustoms them to it and caters for the instinct to browse. A few concentrates can be introduced when they have started really eating the hay. Weaning can be at around twelve weeks, by which time they will usually have learnt to drink water from a bucket.

A good goat will lactate into a second year, with the yield gradually declining. As the milk from one goat was sufficient for our needs, we had our goats – Florence, a British Saanen, and Bilberry, a British Alpine – mated on alternate years so that as one yield was declining, the other took its place.

Health

Reference has already been made to the importance of only buying animals from CAE-free herds. If goats are well fed and have proper housing and management, they are likely to remain healthy and stress free, but they can be affected by diseases found in other ruminants. The relationship between the goats and their owner is important: it is important, for example, to learn the specific characteristics of individual goats. They are intelligent and respond well to kind attention, and will answer to their names when called. This is useful in spotting conditions that are out of the ordinary. Suspicious symptoms that might require veterinary attention include lack of appetite, poor coat condition, wounds, lameness, sore teats, and loose droppings (they should be round and firm like a rabbits).

Vaccinations
At the beginning of the season, before they go out to grass, all the goats should have an anti-clostridial vaccination, as is the case with sheep.

Worming
Goats, like all livestock, need regular worming, although rotation of pasture should not be neglected. The most common form in which a vermifuge is administered is as a drench or liquid. With the aid of a helper, the goat's head is held up while the drench is inserted through a tube via the side of the mouth. Injections are also available for this purpose. Vermifuges are needed at least twice a year and possibly four times, depending on the use of the land.

Horn Disbudding
Although it is normal for Angora and Pygmy goats to be left with horns, dairy goats are normally disbudded for safety reasons. Kids should only be disbudded during the first week of life, a procedure that must only be carried out by a vet.

Foot Trimming
Like sheep, all goats need their feet checked regularly, so that overgrown nails (cleats) can be trimmed, as well as ensuring that there is no foot rot infection.

Mastitis
Heavy milkers are susceptible to the bacterial infection mastitis, that causes lumps in the udder and

Goat kids in France being reared in communal pens until they are ready to go outside. Note the slatted dividing pens so they can see each other, and the milk multi-feeders on the left.

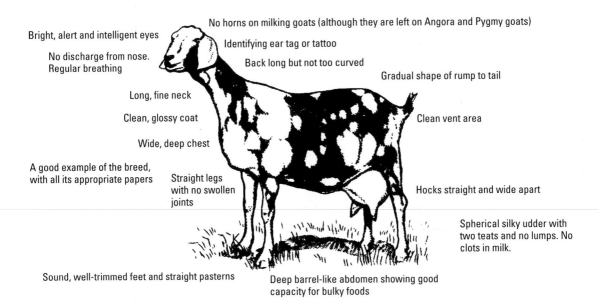

Bright, alert and intelligent eyes

No discharge from nose. Regular breathing

No horns on milking goats (although they are left on Angora and Pygmy goats)

Identifying ear tag or tattoo

Back long but not too curved

Gradual shape of rump to tail

Long, fine neck

Clean, glossy coat

Clean vent area

Wide, deep chest

A good example of the breed, with all its appropriate papers

Straight legs with no swollen joints

Hocks straight and wide apart

Spherical silky udder with two teats and no lumps. No clots in milk.

Sound, well-trimmed feet and straight pasterns

Deep barrel-like abdomen showing good capacity for bulky foods

What to look for in a healthy goat.

Goats are intelligent and sensitive animals, and respond well to good care. These are Golden Guernsey goats.

Angora goats are not disbudded, and their horns are allowed to grow.

clotting in the milk. It must be treated with antibiotics, and the milk should be discarded for three days after the last treatment.

Bloat
Too much green food, such as kale, can cause bloat or a build-up of gases in the rumen. An excess of any one type of food, including concentrates, should be avoided.

External Parasites
Like all stock, goats are liable to pick up external parasites such as fleas, lice and ticks, and they should be dealt with by the use of a proprietary product.

Milking

Goats should be milked twice a day, morning and evening. For a household supply, hand milking is the usual practice, although small milking machines are available for cows, goats and milk sheep. Where milk is sold, the milking, filtering and pasteurizing procedures will all be mechanized. The following instructions are for those who are producing milk for home consumption only.

The udder should first be wiped, and the foremilk or first milk squeezed into a 'strip cup' for examination. A 'strip cup' is a special cup manufactured with a black internal surface so that close examination for clots, blood spots and other indications of mastitis and infection is possible. The

first few squirts will also contain bacteria and so should be discarded. Once this is done, the rest of the milking can take place as quickly as possible. For hand milking, a stainless steel bucket is the best container to use, for it is easy to clean and sterilize. Once milking is complete, it is a good idea to dip the teats in a proprietary solution that will help to protect the teats against infection, as well as keep them smooth and soft.

The milk should be filtered and cooled immediately. Purpose-made filter units are readily available. The milk is then cooled and put into cartons. These are sealed and put in the refrigerator. Goat's milk with its small butterfat particles also freezes well, so cartons can be dated and stored for subsequent use.

If pasteurization is required, the milk can be heated in a pan or purpose-made pasteurizer. Remove from the heat just before it begins to simmer at the surface, allow it to cool slightly, then pour it into the cartons and put in the refrigerator until required.

Home Dairying

Having a dairy animal for a domestic milk supply makes possible the production of a range of dairy products. It must be emphasized again, that for commercial production, the premises must be registered and inspected so that all the requirements of the dairying, food safety and labelling regulations are met. The milk must also be tested on a regular

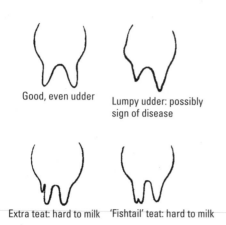

Good, even udder   Lumpy udder: possibly
sign of disease

Extra teat: hard to milk 'Fishtail' teat: hard to milk

Conformation of the udder.

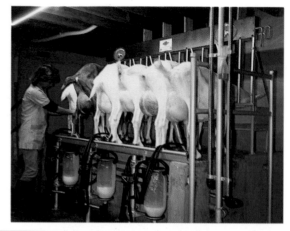

Machine milking for goats or sheep. Photo: Fullwood

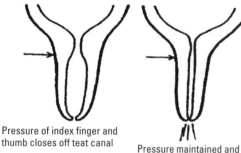

Pressure of index finger and
thumb closes off teat canal

Pressure maintained and
remaining fingers close
against hand, forcing milk out

Milking sequence.

Detachable examination dish

Spring attachment
for dish

Foremilk examination cup.

basis. The following information is for those who wish to produce dairy products for home consumption only. Scrupulous attention should be paid to cleanliness, with all utensils sterilized before use.

Cream

If milk is left to settle in cool conditions, the cream will rise to the surface. From here, it can be skimmed off with a skimming ladle. Goat's milk has smaller butterfat particles than cow's milk, and the cream rises to a lesser extent. An alternative is to use a cream separator that separates the cream by means of centrifugal force. However, it is probably not worth it for a relatively small amount of milk.

Butter

Small, table-top butter churns are available that are operated by hand, or are electrically driven. Alternatively, a kitchen mixer can be used. The cream is left to ripen in the refrigerator for a few days, and is then put in the churn. It is then rotated at a fast speed until 'breaking point' is reached: this is the point at which butterfat droplets coalesce and merge together into a mass. The mass is washed to remove the milk and then 'worked' to remove all traces of liquid. Scotch hands were traditionally used for this. The butter can be salted if required, and put into appropriate containers.

Yoghurt

The easiest way to make yoghurt is to use a thermos flask. Pasteurize the milk by heating it to 82°C (180°F), and keep it at that temperature for a few minutes. Purpose-made dairy thermometers are available. Cool the milk to 43°C (110°F) then stir in a 'starter', a live culture of lactic acid-producing bacteria, available from dairying suppliers. Pour the milk with the blended starter into a sterilized thermos flask and leave overnight. The following day, the yoghurt will be ready. If necessary it can be strained to remove some of the liquid, leaving the curds to form a thicker yoghurt. Milk with a relatively high butterfat content is needed, otherwise the yoghurt will be too watery. Where goat's milk is used, that from Nubian goats is best, otherwise powdered milk may be necessary to thicken it.

Soft Cheese

To make cheese a different starter culture is needed, as well as vegetarian rennet to coagulate the milk. Again, they are both available from specialist suppliers. Pasteurize 1 litre (around 2 pints) of milk, as indicated earlier, then cool to 30°C (86°F) for cow's milk; 2°C less for goat's or ewe's milk. Add the starter culture and stir in thoroughly. Leave to ripen for ten minutes. Add three drops of rennet to a tablespoonful of previously boiled and cooled water. Stir into the milk, cover and leave until it has coagulated and separated into curds and whey. Strain the cheese and add a little salt to taste. Pepper or garlic can also be added if these are liked. Form into a round, or put into a container and put it in the refrigerator. It should be eaten within a few days.

Hard Cheese

Making a hard or pressed cheese is more complicated, and is not worth it for less than 5 litres (approximately 1 gallon) of milk. Some extra equipment will be needed, including a large stainless steel pan or vat, a cheese mould or form in which to put the curds, and a cheese press for exerting pressure on the cheese. Pasteurize the milk as before, and cool to 32°C (90°F). Add the starter culture and leave for half an hour. Mix 1 teaspoonful of rennet with 3 teaspoonfuls of previously boiled and cooled water and stir into the milk. Leave for 30–40 minutes until it has set into a firm curd that does not leave a milk stain on the back of the finger.

Cut the curd into strips with a long knife, then into

Cutting the curd, one of the stages in making a pressed cheese.

squares. Loosen the curd around the edges of the pan and then cut diagonally so that it is in pieces. Leave until whey shows above the curds, then increase the heat slowly to 38°C (100°F) over a period of half an hour. Stir the curds from time to time, then turn off the heat and leave them to settle for another 30 minutes.

Drain the curds into a muslin cloth and tie up in a bundle. Open the cloth after 15 minutes and cut the now solid curd into four thick slices. Stack them, one on top of another, and leave for a quarter of an hour, then rearrange the order of the slices, with the middle ones on the outside. Leave for a further fifteen minutes, then break the curd up into pieces the size of a nutmeg, and add salt to taste. Place the curds into a cheese mould or form lined with enough muslin to cover the cheese, then place it in a cheese press and exert pressure on the wooden 'follower' placed on top. The next day, turn the cheese upside down and press again, leaving it until the next day. Remove from the press and muslin, and leave to dry for two to three days. It can then be coated with cheese wax and stored in a cool room. Turn the cheese once a week, and it will have ripened after five weeks. Leave for longer if a more mature flavour is preferred.

17 Cattle

O Mary, go and call the cattle home across the sands of Dee. (Charles Kingsley, 1858)

If there is sufficient pasture, it is possible to keep cattle. Registration is required, as well as compliance with all the welfare and identification legislation. Procedures vary in different countries, but in Britain, the Animal Health Office and the local agriculture office must be contacted. As with other farm animals, it is necessary to keep herd records, a medicine and movement book, and to comply with the ear tagging regulations. Each cow or bullock also needs a passport that should accompany it when it is moved, and individual files must be kept for each animal. If you already have animals insured, the cover can be extended. If not, you will need public liability that includes cover for zoonoses. There are insurance brokers who specialize in livestock cover.

When buying cattle, the safest route is to choose pedigree stock and purchase direct from a breeder. This may cost a little more, but if animals are to be sold in the future, they will command a far better price. Pedigree animals also come with a warranty.

What to buy depends on your intentions. If you wish to raise cattle for meat, buy weaned and castrated bull calves; if you intend to breed, buy heifer calves or older heifers ready to breed at around eighteen months old, or even in-calf heifers. Perhaps the best bet is to purchase an in-calf cow with a calf at foot – and if this is done, try to obtain information on her breeding history. Realistically, cows can produce for about twelve years.

Breeds

If the decision is made to keep pure breeds (and this is certainly the best for smallholders), watch out for 'improved' versions! These have been cross-bred in North America and re-imported into Britain. They look similar, but are invariably larger, and may have lost some of their original characteristics; this process is known as 'introgression'. The breed societies can provide further information.

Jersey cow and calf. This breed produces high quality, creamy milk and is popular with smallholders.

Cattle can be classified as dairy, dual-purpose or beef breeds. A house cow that provides milk for the family does not have to be a prolific producer, and a smaller Channel Island breed or dual-purpose cow may be a more practical proposition.

Dairy Breeds

Jersey: The most popular dairy cow for the smallholder is the gentle Jersey. It produces creamy milk, high in butterfat, that is ideal for home dairying activities.

Guernsey: Another Channel Islands' breed, the Guernsey is also good for dairying activities. It has been 'upgraded', but original breed types can still be found.

Other dairying breeds are the Ayrshire and British Friesian. These are larger cows that have both been developed through the North American Holstein so that original types are hard to find, particularly in the case of the Ayrshire.

Dual-Purpose Breeds

Dexter: The Dexter is a small breed that is popular with smallholders. The beef is well regarded, and the smaller joints are popular with customers. The cows

A Dexter cow and calf in their exercise yard outside their house. From here they have access to pasture.

are thrifty, and work well as house cows. Unfortunately the breed has a lethal gene that sometimes results in unviable 'bulldog' calves. This is less of a problem these days, but avoid the short-legged animals.

Dairy Shorthorn: A hardy cow that does well on poorer pasture. Look for the original type rather than improved versions.

Kerry: An old Irish breed suitable for the smallholder, but it is not available in large numbers.

Gloucester: An attractive red and white cow. It has become more of a beef animal today, but can still be a good house cow. Its milk was the original source for Double Gloucester cheese.

Another dual-purpose breed worth considering is the **Irish Moiled**. It is a rare breed that is hardy and available as a dairy or beef type. **Shetland** cattle are also hardy little cows that do well on poor grazing. **Red Polls** are hornless, hardy cows that produce early maturing beef. They make good mothers and house cows, if you can still find a milky strain.

Beef Cattle

If cattle are to be raised for beef in Britain and the European Union, BSE precautions require animals to be slaughtered before thirty months. With some of the slower maturing breeds, it may be necessary to provide extra feeding to get them ready by this deadline. Early maturing breeds should finish between 18–24 months.

Aberdeen Angus: A black- or red-polled (hornless) animal, but so developed that there are few original

types still available. The smaller, early-maturing types are the most applicable for smallholders.

Beef Shorthorn: A good beef producer, and there are some British herds that include original stock without imported bloodlines.

British White: A minority, polled breed, that is being graded up and becoming more of a beef breed.

Galloway: Another polled breed and a very hardy cow that copes well with poor grazing. It is good for hilly and wet areas.

Hereford: This traditional early-maturing breed has been exported and developed all over the world, but there are numbers of the original type still available in Britain. It is a good beef producer.

Highland: Suitable for harsh weather conditions, the Highland is a good beef producer, but is slow maturing. The Luing is a recent breed developed by crossing the Beef Shorthorn bull with Highland cows. It does well on poor land, but is available only in Scotland.

British White calf.

The long-haired Highland breed of Scotland is one of the hardiest breeds of cattle.

Longhorn cattle, an ancient breed that has many active supporters in Britain.

Lincoln Red: This is a large breed with a good temperament, but there are few still available.

Longhorn: A large and docile minority breed with large horns, the Longhorn is an ancient breed that produces good beef.

North Devon: Originally a dual-purpose cow but now a beef breed, this is a large animal that was often used to provide draught oxen. Devon cattle are hardy and early-maturing animals producing quality beef.

Sussex: This is another large, docile breed that matures early from poor land.

Welsh Black: An old Celtic breed, the Welsh Black was originally dual purpose, but has now been selected for beef, with good growth rates. My parents always had them in the days when they were still used for milk. Wales also has the descendants of other *gwartheg hynafol* (ancient cattle); they come in a variety of colours.

White Park: The oldest recorded breed in Britain, as of 2,000 years ago. They are white animals with black points, very hardy, and good beef producers from poor forage.

Housing and Pasture

Few people realize that the cow's ancestors were originally woodland animals: consequently they do best if they have trees, hedges or other shade areas in their pasture. One animal needs 0.4 hectare (1 acre) of good grazing land, although beef cattle can be kept at three animals to 0.8 hectare (2 acres). Cattle are strong and heavy, so a sturdy 1.2m (4ft) high fence is needed for dairy cattle, while 1.5m (4ft 6in) is better for bullocks. If this is a first experience of keeping cattle, do not be tempted to over-stock.

A good field shelter with a hayrack is necessary. If the cows are to be brought in for the winter, they will need a waterproof building with good light and ventilation, as well as an exercise yard. Bedding needs to be topped up with straw to keep it clean and dry. Periodically, it will all need to be cleared out and replaced, and the old straw composted.

Feeding

The most important source of food is grass, and this is why most of the dairy herds are in areas where there is abundant rainfall, and therefore green pastures. Grass is only available from spring to autumn; once it has stopped growing, and winter has set in, the grass must be fed in a dried form as hay, or as silage or 'pickled grass', frequently fed to cattle if the scale warrants it. As already discussed, it is difficult and unsafe to try and produce it on a small scale, which is why most smallholders use hay only.

Fodder crops such as kale, cabbage and sugar beet are grown for cattle feeding.

Water is essential at all times. The best way of supplying this is to have a tank in the field that refills automatically from the mains, controlled by a ball valve.

A dairy cow needs a maintenance ration – an adequate supply of food to keep her in good condition when she is not in milk – and once she starts her lactation she will need a production ration, which is the maintenance ration plus an extra amount, depending upon her level of production. In commercial herds, this is estimated accurately so that the ratio of feed to milk produced is as cost effective as possible. On a small scale, with just one or two milking animals, the owner is generally more relaxed about it. Individual cows do, of course, have different requirements, and a large Friesian will consume more than a small Jersey. The genetic make-up of the animal also has a bearing on this.

The amount of food that any cow will eat is limited by its digestive capacity. If it ate only bulky food such as hay, it would be replete once it had

Organically reared beef cattle receiving some winter fodder.

reached its capacity; however, its nutritional needs in relation to the milk it produces might still not be satisfied. For this reason, it is important to control the amount of bulk given, and to ensure that concentrates are given in the right balance. This balance will depend upon the level of production, in that the higher the volume of milk produced, the greater the amount of concentrates required, in relation to bulk foods. It is also important to feed a proportion of hay in the mornings before the cows are let out to pasture in the early spring. At this time, the new grass has its most laxative effect, leading almost certainly to scouring. Barley straw can be used instead of hay for this purpose, as well as for supplementing the grass later in the season. The grass intake itself can be restricted by the use of electric fencing. It is useful to remember that one normal-sized bale of hay weighs around 20kg (44lb). Big bales are available, but these are generally used by farmers who have the means of moving them mechanically.

Concentrates are normally mixtures of high-energy grains or grain-based compound feeds; the latter are available from feed suppliers as proprietary dairy cattle rations. They contain balanced nutrients and minerals, including the important mineral magnesium, often in short supply in new spring grass. A deficiency of this can lead to a condition known as hypomagnesaemia or 'grass staggers'.

Organic concentrates are available. Typical organic dairy cubes have a protein level of 15–16 per cent, while organic beef cubes are normally 14 per cent protein.

Breeding

On a small scale, the keeping of a bull is not a practicable proposition. Bulls can be dangerous, they need special penning and expert handling, and are unlikely to pay for their keep in a small enterprise. With the availability of artificial insemination, there is no need to have one.

Bulling

The period of heat is known as 'bulling'. It is usually indicated by fretful, restless behaviour, an almost continuous mooing, and either trying to mount other animals or standing still when other cows try to mount her. This period is not without its dangers. My husband was once bending over, clearing out a water tank in a field when a bulling Jersey heifer landed full square on his shoulders! Fortunately he was able to escape her amorous attentions without harm. Another sign of 'bulling' is a slight colourless discharge from the vulva.

Heifers are normally served at 15–18 months old. Pregnancy lasts for about nine and a half months, although there may be a few days either way. During this time the cow needs her normal maintenance ration, together with a ration to cater for her current level of milk production. This will gradually taper off, until about two months before the calf is born, she should be dried off. This must be done carefully to avoid mastitis. The principle is to milk less frequently, because the less that is drawn out, the less is produced.

Calving

About four weeks before calving, the udder will begin to fill out, and later signs immediately before the birth are a slackening of the muscles on either side of the tail, while the vulva itself enlarges. Calving can take place outside, particularly if the weather is fine, and there is usually less danger of infection in these conditions. Shade and wind protection should be provided. If the cow is brought inside, she should be put in a stall that has been well scrubbed out and supplied with clean, fresh straw.

Most calvings are straightforward, but the telephone number of the vet should be to hand in case of emergency. Again, the importance of having attended a practical course of instruction cannot be over-stressed.

For the first few days, it is essential for the calf to suckle its mother. In this way it is able to have the colostrum or first milk. It has a good start in life from this, and is protected against infection until it is able to produce its own antibodies. Bull calves that are being raised for meat will need to be castrated, as detailed in the sheep section, while all calves must be tagged or tattooed for identification. Any disbudding to prevent horns growing must be carried out by a vet.

From the fifth day onwards, the calf must be separated from the mother and put in its own pen. Alternatively, several calves can be penned together, and experience shows that they respond far better to separation from the mother if they have other company.

The calf will need to have a milk substitute twice a day up to the age of six weeks, and this needs to be made up according to the manufacturer's instructions. It will also need to be taught to drink rather than suck. The way to do this is to have a bucket of milk ready, then give the calf your finger to suck. Gradually lower your hand until it is in the bucket, and as soon as the calf's nose reaches the milk, it takes milk in as well as your finger. Gently remove your hand once it has started to drink by itself. It will snort and sneeze for a while, until it gets used to this

Calves soon learn to feed from a bucket when they are weaned.

Calves, such as these young Jerseys, are less likely to experience health problems if they are able to go outside as soon as possible.

new way of drinking, but is usually too greedy to stop drinking in order to find your finger again. Hay, fresh clean water and calf concentrates should be available from one week onwards.

From the age of six weeks onwards the calf can be weaned from milk and fed hay, fresh clean water and a proprietary calf weaner ration. During the next few weeks, the amounts are increased from 85–115g (3–4oz) a day to about 1.3kg (3lb) by the age of nine weeks.

When the weather is mild the calves can be allowed out to graze on young, fresh grass, but the grazing must be controlled, and should not exceed a couple of hours for the first week or two. This is to avoid scouring, which can be brought on by an excess of lush grass in the spring. Reference has already been made to the need for making hay available before going out.

Machine Milking

Details of hand milking for a domestic supply of milk were given in the previous chapter. The principle is the same for cows, except that they have four teats, not two. There are individual milking machines for cows, goats or milk sheep, with teat clusters of varying sizes, as appropriate.

It is unlikely that a smallholder will go in for milk sales because the regulations that apply would make a small enterprise non-viable. In Britain, for example, it

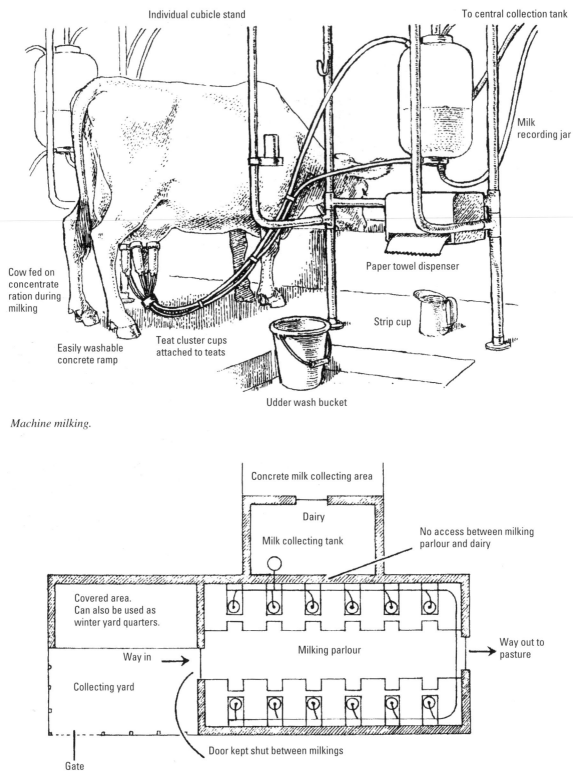

Individual cubicle stand

To central collection tank

Milk recording jar

Paper towel dispenser

Cow fed on concentrate ration during milking

Strip cup

Easily washable concrete ramp

Teat cluster cups attached to teats

Udder wash bucket

Machine milking.

Concrete milk collecting area

Dairy

Milk collecting tank

No access between milking parlour and dairy

Covered area. Can also be used as winter yard quarters.

Way in

Way out to pasture

Milking parlour

Collecting yard

Door kept shut between milkings

Gate

Traditional farm buildings adapted for a small dairy herd.

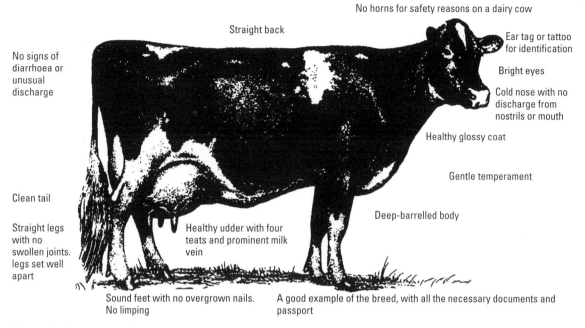

No horns for safety reasons on a dairy cow

Straight back

Ear tag or tattoo for identification

Bright eyes

No signs of diarrhoea or unusual discharge

Cold nose with no discharge from nostrils or mouth

Healthy glossy coat

Gentle temperament

Clean tail

Deep-barrelled body

Straight legs with no swollen joints. legs set well apart

Healthy udder with four teats and prominent milk vein

Sound feet with no overgrown nails. No limping

A good example of the breed, with all the necessary documents and passport

What to look for in a healthy cow.

is necessary to purchase a 'milk quota' and meet all the requirement of dairying regulations, with milking parlours, equipment and procedures being subject to inspection. The USA has similar regulations.

The principle of machine milking is that a vacuum created by a pump is produced in the cups enclosing the teats, and this in turn produces a sucking action. Regular checking is necessary to ensure that the pressure is correct, otherwise the cups can be pulled up too high, causing damage to the teats. As with hand milking, the first necessity is to ensure that the udder is clean and free from mud or other contaminants. The foremilk is also discarded before the teat clusters are attached. Milk is then extracted from the udder, and transferred via a pipeline into a receiving receptacle. In a commercial milking parlour, the milk goes into a recording glass jar and then to a central refrigerated tank or vat to await collection by a bulk milk tanker. The teat clusters are then removed, and the teats are dipped individually in a container of proprietary solution to protect the cow from mastitis, as well as keeping them soft, smooth and in good condition.

Health

Following the tragedy of bovine spongiform encephalopathy (BSE), the regulations for keeping cattle in Britain are now the most stringent in the world. It is thought that BSE was caused by including animal protein from scrapie-infected sheep in calf rations: the scrapie pathogen mutated and successfully crossed the species barrier. Many areas of the world are still including animal protein in cattle feeds, a recipe for disaster where naturally vegetarian ruminants are concerned.

Notifiable Diseases

There are certain notifiable diseases in relation to cattle, and these include cattle plague, anthrax, foot and mouth disease, pleuro-pneumonia, and rabies. Any sudden and unexplained illness must be reported to the vet immediately. He will also advise on what vaccinations are appropriate.

Vaccinations

If cattle are well looked after, they should remain free of infections, but like sheep and goats, they should be vaccinated against the range of clostridial diseases. Protection in the form of combined vaccines are available. These are normally given before the cattle go out on grass, with a booster a few weeks later. Thereafter an annual vaccination is needed, unless the vet advises more frequent protection.

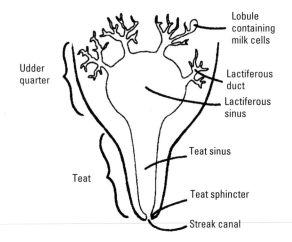

Udder quarter

Teat

Lobule containing milk cells

Lactiferous duct

Lactiferous sinus

Teat sinus

Teat sphincter

Streak canal

The udder.

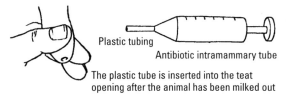

Plastic tubing

Antibiotic intramammary tube

The plastic tube is inserted into the teat opening after the animal has been milked out

Treating mastitis of the udder.

Tuberculosis Testing

Cows are regularly tested for tuberculosis under the auspices of the agricultural authorities. A healthy cow is alert and active with clear bright eyes, a moist pink lining to the mouth, lips and nostrils, soft flexible skin, and glossy hair. To maintain their health, they should have access to clean pasture that is rotated regularly.

Internal Worms

A problem with internal worms is often indicated by a poor 'staring' coat or a persistent cough. Every effort should be made to provide new, clean pasture on a regular basis. A suitable anthelmintic from the vet can be administered.

External Parasites

External parasites such as flies, lice, mites and warble fly can cause problems. Providing shade in hot weather and making every effort to avoid scouring are essential. Stockholm tar, iodine, Cyromazine and Deltamethrin are effective.

Scouring

Scouring can be the result of too much lush grass. An excess of concentrates in relation to bulk in the form of hay can also cause laminitis and lameness. Scouring, when accompanied by poor skin condition, may also indicate internal worms. It is best treated by giving warm water to drink and hay only to eat for a day. If it persists, or if it is apparent in calves, the cause is likely to be bacterial and veterinary advice is essential. *E. coli* and the salmonella species can affect humans as well as animals.

Mastitis

Reference has already been made to the importance of preventing mastitis. When it does occur, it will need an antibiotic infusion administered into the teat aperture of the affected quarter. It is not difficult to do, but should be demonstrated by a vet or other experienced person first. In a dairy animal the milk should continue to be drawn from the affected quarter and discarded until a few days after treatment has finished. Organic farmers have reported success with homoeopathic remedies for mastitis.

18 Fleeces

Silver threads among the gold.

Some animals are raised for their high quality fleeces; these include alpacas and Angora goats. Llamas have coarser fleeces, but have a following amongst those interested in wool crafts; they are also used as pack animals. Alpacas and llamas are camelids from the Andean region of South America. Domesticated and bred by the Incas for thousands of years, alpacas are renowned for the softness and durability of their fleeces.

Camelids deposit dung in selected spots, making clearing more straightforward; this also helps to conserve pasture. They are herd animals and do not thrive in isolation; they are also ruminants and chew the cud, like sheep and cattle. However, like goats, they are browsers rather than grazers, relishing a wide range of plants and poor quality vegetation. They have soft-padded feet with only two toes, rather than hooves, and are much less destructive of pasture as a result. They need to have their feet trimmed about once every few months. Other tasks include worming and vaccinations, as for sheep and goats, to prevent health problems. They live for fifteen to twenty-five years.

Family groups of alpacas out on pasture.

Alpacas

The average adult has a height of about 1m (39in) at the shoulder, and weighs around 70kg (154lb). Average pasture density is twelve to fifteen alpacas per hectare (five to six per acre). They do not require any special buildings, beyond the need to have sound shelter from the weather. They adapt well to being trained to the halter and are easy to confine, not being jumpers. Ordinary sheep netting is sufficient.

Alpaca farming is an expensive activity to get into, and for livestock as valuable as these, adequate insurance cover is a must. Most people in the USA and Europe have gone into it in a small way to begin with, concentrating on breeding and building up a herd. There are those who will be interested in keeping a few alpacas on a small acreage, as part of a crafts enterprise. This is an area currently growing in the USA, particularly where coloured fleeces are concerned, and there are many enterprises now selling stock, wool, garments and decorative textiles. As with any activity, planning, good research and advice are essential; but there seems to be a bright future for alpacas.

Breeds
There are different types of alpaca, differentiated according to their fleece.

Huacaya: This is the most common type, and is generally considered to be the most hardy. It has fairly dense body wool, extending down to the legs. On the forehead is a 'bonnet' of wool, while the cheeks are definitely 'mutton-chops'. The wool is generally more wavy, making the fibre popular with spinners. It is available in a range of eight colours and twenty-two hues of white, black, grey, red, coffee, caramel, fawn and piebald.

Suri: Much rarer than the Huacaya, the Suri is predominantly white or fawn, although there is increasing interest in the USA in breeding coloured animals. The fleece is lustrous, with a silky feel to the staples that hang in long, fine locks with little or no crimp (waviness). The Suri tends to do better in more temperate climates, hence its current popularity in Australia and the milder areas of the USA.

Chili: A relatively new designation coined in the USA to describe alpacas that are halfway between Huacayas and Suris. The fleece is more open, less dense and softer than that of the former, yet it does not have the well-defined, straight staples of the latter. Chilis have a long, straight fringe on the forehead, making them easy to recognize.

Housing
Alpacas can cope well with cold weather, as long as they have shelter from the rain. Simple open-fronted field shelters or barns are suitable. These should face away from the prevailing winds and have clean bedding straw. A hayrack in the shelter can be kept topped up with hay. A feeder for the concentrate ration and a water trough are the only other necessary items.

They are docile and easy to handle, and can be halter-trained. There are suppliers who produce halters and other accessories for alpacas and llamas. Alpacas and llamas can be grazed with sheep, goats or cattle, but extra care needs to be taken with routine worming and rotation of pasture.

Feeding
Alpacas can suffer from vitamin and mineral deficiencies, so it is important to ensure that they

The stately llama is a first-rate pack animal; it also protects poultry against foxes.

have a balanced diet. In addition to their browsing activities, they are normally given a concentrate feed that has been specially formulated for them. The amount will vary, depending on the availability and quality of the grass. In winter, for example, each animal will need 250–500g (9–18oz) of concentrates a day, as well as hay. It has been estimated that alpacas eat around a tenth of the food consumed by a horse.

Breeding

Alpacas are ready for breeding at eighteen months, at which time the females will weigh around 45kg (99lb). Males should be two to three years of age. The average gestation period is 335 days. Birthing is generally trouble free, with alpacas often giving birth during the day. The young are weaned at six months of age. A baby is called a 'cria' and weighs around 6–7kg (13–15lb) at birth; it will immediately struggle to stand. There is usually only one baby, with twins comparatively rare. The relatively long pregnancy and general lack of multi-births is something of a drawback, particularly as there is a shortage of breeding stock. Conversely, the alpaca is an induced ovulator in that breeding can take place all year round, without there being a limited breeding period. Commercially, re-mating takes place fourteen days after birth.

Llamas

Llamas are less expensive than alpacas. They require the same type of housing and management, and can be stocked at around ten per hectare (four per acre) on good pasture. Normal stock fencing is suitable, but it is advisable to have an extra strand (not barbed wire) above it. They are larger and heavier than alpacas, weighing 130–160kg (285–350lb) when adult. Most people who keep llamas use them as pack animals, and they are well suited to this, for they are strong and have been bred to the task. Llama trekking is a popular holiday activity in Britain, where the llamas carry the food and supplies for those enjoying an outdoor holiday.

Llamas have proved themselves to be good guardians of sheep, especially at lambing time. They also have a good reputation for protecting poultry from foxes. Some people have them as 'companion' animals, to provide company for an otherwise solitary animal such as a donkey.

The gestation period is around 345 days, with re-mating taking place two to three weeks later. As with alpacas, a single cria or calf is usual, and weaning takes place at the age of six months.

Angora Goats

Angora goats produce mohair, a fine lustrous fibre

Angora goat, a breed renowned for its quality mohair.

much in demand in the quality textile field. The pure white Angora is most attractive, with long ringlets of lustrous hair. An ancient breed recorded in the Bible, the Angora comes from Turkey and has been introduced to many parts of the world, including Britain, the USA, Australia, New Zealand and South Africa, all of which have contributed to the standard of the animal today. It is customary to leave horns on Angora goats.

Their management is similar to that of sheep, including the need for regular foot trimming every six to seven weeks. They prefer longer grass and are happy sharing poorer pasture with horses. They can be followed by sheep, but as they share the same worms, a worming programme is essential. They are easily contained by normal sheep netting. Care of the living fleece is vital; for instance, fleece quality can be reduced by exposure to persistent rain so some shelter is essential, particularly from winter storms – although Angoras are from a dry region, they have adapted well to damper climates. It is also important to feed hay from a low hayrack so that seeds do not drop into the fleece. External parasites must be controlled.

Angoras breed seasonally, and mate in the autumn. Gestation is around 150 days, and kidding rate is normally about 150 per cent in a flock. Unwanted buck kids can be castrated, as detailed earlier, and kept for mohair production. Twins are more common with older does. They mature slowly, and are kidded when approaching their second year. Kids are normally left to suckle their dams, and are weaned by the midsummer shearing.

Feeding requires a daily supplement of seaweed meal and access to a cobalt salt block, together with coarse hay fed ad lib for most of the year. Concentrates should be fed according to the doe's level of production. Pregnant and milking does are fed at a similar rate to lowland sheep, and kids should be well fed in their first year; thereafter the amount of concentrates can be reduced. The aim is to supplement the protein, energy and mineral content of grass to meet the Angora goat's needs.

Wool and Fibre

There is sometimes a degree of confusion when referring to different animal hairs: *wool* comes from sheep or Angora rabbits, while *fibre* comes from alpacas, llamas and Angora goats. Some also have special terms: thus Angora rabbits produce 'angora wool', while Angora goats produce 'mohair'. Cashmere comes from Cashmere goats, but it should be noted that there is no specific breed of goat of this

name, it merely refers to goats with a particularly thick secondary coat, a feature often found in mountain goats that need to keep warm. Angora goats have been crossed with feral goats and other breeds to increase the cashmere value. While goats are being graded up to cashmere production in this way, the fibre is referred to as 'cashgora'.

Any fleece is made up of primary or guard hairs, and an undercoat of secondary hairs; the former are coarse and reduce the value of a fleece. Secondary hairs vary in the quantity and degree of fineness, depending on the type and breed of animal. Fineness is measured in microns or the micrometre diameter of each hair, with the finest fibre having the lowest number (1 micron = 0.001mm). The Bradford Count, representing both fibre length and fineness, expresses the theoretical length of yarn that can be spun from 1lb of wool, and is quoted in hanks. Here the opposite applies, namely the higher the number, the finer the quality.

The hairs in a fleece tend to arrange themselves in locks called *staples*, and depending on the type and breed, these will be short, medium or long. They have a certain degree of waviness or *crimp*, again a measurable quality depending on how many waves there are to the inch.

Each secondary hair has microscopic scales that allow it to cling on to other fibres, giving it strength and making spinning possible. *Lustre* is a characteristic of some scales to reflect light, and mohair is often called the 'diamond' fibre for this reason.

Shearing

Shearing is essentially the same for all fibre-producing animals, with obvious variations for the size and type. With sheep, shearing takes place annually and the aim is to part the old wool from the new, cutting off the whole fleece at the new white wool level. The sheep should be dry before shearing takes place and for this reason it is best to pen them under cover overnight. Food should also be withheld that morning, to prevent the fleeces becoming soiled by droppings.

Sheep: Shearing is a skill that can be acquired with tuition and practice. It is important not to nick the animal's skin. Contract shearers are available for larger numbers of animals. Alternatively, shearing can be done by the flock owner. Hand shears are suitable for a small number, but motorized shearing equipment is also available; 12v battery-operated shears are useful where there is no convenient supply of electricity.

Include only clean, dry belly wool with the fleece before folding in the flanks towards the centre.

Turn in the britch end and roll the fleece firmly and neatly towards the neck.

After rolling, part the fleece.

Without twisting, tuck the neck wool firmly into the body of the parted fleece . . .

. . . resulting in a well presented fleece which is firm and secure. Place the tucked-in fleece firmly into the wool sheet. Photos: British Wool Marketing Board

The sheep is cast, leaning slightly to one side. The wool is clipped first from the face, then down the brisket or upper chest to the abdomen, so that the fleece is parted at the belly. One hand ensures that the skin is kept taut and also protects vulnerable areas such as teats and genitals. Once the wool is cleared from the belly, clip the inside legs, ensuring that each leg, in turn, is stretched out while being clipped. Place the sheep on its side and continue with the outside leg. The wool is clipped back along the side, working from the belly to the backbone. When these cuts are complete, shear from the tail upwards, keeping parallel with the backbone. Repeat on the other side, making sure that the sheep does not get up on all four feet. Clip the front legs next, concentrating on the inside first, then working around to the outside, and then joining up with the

Newly shorn sheep awaiting release to their sunny pasture.

Spinning is a popular craft in the countryside.

previous cutting by taking cuts running parallel with the backbone, up to the head. Trim the top of the head and the cheeks until the fleece is completely removed. Release the sheep and lay the fleece on a clean surface. Fold the flanks inwards, then roll the whole fleece up towards the neck end.

Alpacas: Alpaca fibre is similar to cashmere, very fine and with most of the clip being 30 microns or less. Shearing is done annually, in early summer. It normally takes two people to shear; one to restrain while the other does the trimming. The animal remains standing, rather than being cast. The average production is up to 3kg (6.6lb) a year for a female and 5kg (11lb) for a male, but this is a generalization, and there will be variations depending on the type and age. The finest wool of all comes from the first baby clip. Shearing is the same as for sheep, although as there is no lanolin in the wool the shears need to be kept well oiled. The fleece is dense with very few coarse guard hairs, while the partially hollow fibres make it light and warm.

Llamas: Llamas have a double coat, with coarse, outer guard hairs and a soft, downy undercoat that is very fine. The problem is that the guard hairs have to be separated from the undercoat, an option that adds to the cost of processing. The yield of fibre is around 2–3.5kg (4–7lb).

Angoras: Angora goats are sheared twice a year, in midsummer and midwinter. Kids are shorn from six months onwards, and their fleece is the most valuable.

Fibre-processing companies are interested in large quantities only, and small producers need to co-operate and pool their crop. In Britain, for example, the Alpaca Fibre Co-operative buys in fleece from its members at a fixed price per kilo; they then sort and grade the fibre and produce yarn, which is used in a range of high quality products. The mohair producers have formed British Mohair Marketing, which buys in fleeces from members, grades them, and sells on to mohair buyers.

Most sheep's wool in Britain is taken by the

Weaving is a craft that can be carried out on any scale.

British Wool Marketing Board, although coloured wools from minority breeds are frequently sold by the breeders direct to craft customers. The fineness and staple length of sheep's wool varies considerably among the different breeds. Coarse wools, such as Herdwick, are used for carpets and tweeds, while finer ones, such as Merino and Shetland, are used in knitwear. The Longwools produce lustrous fleeces for speciality yarns.

Spinning

Before fibres are spun, they are usually carded or combed to make them lie in the same direction. An exception is cashmere which does not require carding. It is also necessary to decrease the tension on the wheel when spinning fine fibres. During the spinning process, the staples are drawn out and twisted so that the fibres, clinging to each other, form a thread or yarn. Spinning is not difficult to learn, but it takes practice to do it well. A wide range of spinning wheels are available, including some which come as 'ready-to-assemble' packs. Wool is much easier to spin than alpaca and mohair fibre, for these have no lanolin. On a larger scale, contract spinning and weaving services are available, many of these specializing in quality fibres.

Dyeing

The hanks of yarn can either be left in their natural form, or dyed. Naturally coloured fibres obviously require no dyeing. Natural dyes from plant materials from the garden produce colours and shades that are less harsh than chemical dyes. The principle is to extract the dye from the plant by simmering in water. As a general rule, 450g (1lb) of plant material, such as bark, leaves or flowers, will be required for every 450g (1lb) of wool. Once the water has turned a good colour, the plants are strained off, leaving the clear dye liquid. The yarn can be dyed in this, but they will not be colour-fast unless a mordant is used. Oak bark is a natural mordant.

A mordant creates an affinity between the material to be dyed and the dye itself. They are available from fibrecraft suppliers, and include compounds such as ferrous sulphate, copper sulphate and oxalic acid. Some are poisonous and should always be kept in sealed, labelled containers, well out of reach of children. Always wear rubber gloves when handling dyes, and keep any utensil used for dyeing only. As a general rule, 7–14g (0.25–0.5oz) of mordant is needed for each 450g (1lb) of wool, but this is open to experimentation, depending upon how deep or pale a colour is required.

The mordant is dissolved in water and heated until simmering, but it is important not to boil the water otherwise the wool becomes matted. Leave the wool hanks in the mordant for about half an hour, stirring occasionally to ensure even distribution, then remove and wring gently. They can either be dried, and dyed at a later date, or dyed immediately. To dye the wool, heat the dye water until it is simmering gently. Place the wool in it until the required colour is obtained; then squeeze the wool hanks gently and hang them up to dry. They are then ready to be knitted or woven as desired.

There are many practical courses available for learning the arts of spinning, dyeing and weaving. They also include more specialized activities such as felt production.

19 Exotic Species

Start small and test the waters.

In recent years, smallholders and part-time farmers have been at the forefront of diversification within rural communities. They have been prepared to look at new options and ideas, including many that may be regarded as exotic. Caution is always necessary, however, for it is easy to be beguiled into thinking that large or quick returns are possible with an enterprise. Starting small and 'testing the waters' is always a good policy to follow, while catering for local demands is the best plan of all.

Emus

The emu, *Dromicelus novae-hollandiae,* is a member of the same ratite family of flightless birds as the ostrich. It has soft, drooping brown feathers, and lays greenish-black eggs; it is hardy, and has waterproof plumage. Like ostriches, emus are regarded as 'wild animals', and to keep them in Britain a permit is required. It is also necessary to display a warning notice on the fencing, to the effect that they are potentially dangerous.

Emus are monogamous (unlike the polygamous ostrich), and will bond for life. They are therefore kept as pairs with a recommended pen size of 12.2 × 36.5m (40 × 120ft) for each breeding pair. Fencing of at least 1.8m (6ft) is required, while an open-fronted barn is suitable as shelter. Emus are around 1.5m (5ft) tall, and females breed at two years old. Around twelve to fourteen eggs are laid in the first season, increasing to a maximum of around fifty for subsequent seasons, although thirty to forty is normal. In the wild, it is the male that sits on the eggs. The eggs are bottle green, almost black, and are laid very early in the season, from early winter onwards and normally in the evening. A shady corner in the shelter is suitable for nesting, and really clean, regularly replaced nesting material such as chopped straw is important in ensuring that eggs are kept as clean as possible.

Optimum conditions for hatching emu eggs involve a temperature of 36°C (97°F) and a relative humidity of 25–28 per cent for days 1 to 46; then from days 47–51 the temperature is reduced to 35.5°C (96°F), while humidity is raised to 75 per cent. Heated brooder conditions are required for the first few weeks, gradually reducing the temperature as the chicks grow. They can be fed on ratite chick crumbs that are available from specialist suppliers. Once off heat, they can go into a sheltered pen. At six months they are very hardy and can be fed on ratite pellets. At around fifteen months onwards, the young emus reach sexual maturity and begin to make their distinctive noises, the males grunting and the females drumming. It is also at this stage that the 'life pairing' is established. The male is leg-banded for easy identification, because the sexes are difficult to tell apart. (Micro-chipping as a means of identification of all the birds is recommended.)

Emus are kept for their meat, leather and oil, while the eggshells also have a potential within the craft industry where their dark green characteristic comes into its own. The colour extends to below the outer shell surface, a feature unique to emu eggs, so that interesting images can be carved into them. There is a long tradition of emu egg carving in Australia, with some superb examples of 'Dreaming' legends being incorporated by the Aboriginal craftsmen.

Emus are monogamous and will bond for life.

Carved emu eggs at a show in Missouri.

Rheas

The rhea, *Rhea americana*, comes from South America and, like the ostrich and emu, is part of the ratite family. They are similar to emus in size, being 1.5m (5ft) high when fully grown. They graze on grasses and broad-leaved plants, but also feed on insects, amphibians and small animals.

Farmed rheas are raised for meat, oil, leather and, to a limited extent, their feathers. They are also popular attractions at farm parks; they do not require a permit. They can be fed formulated pellets to complement their grazing and browsing. Adult birds can be kept at twenty-five birds per hectare (twenty per acre). Paddocks should be at least 70m (230ft) long so that rheas can run, for this is part of their nature. Fencing needs to be 1.5m (5ft) high for young birds, and 1.7m (5.6ft) for adults. Electric fencing and barbed wire are unsuitable. Although very hardy, rheas need a shelter that is closed in on three sides, with a facility to close the remaining side if required. The shelter should be at least 2.5m (8ft) high, with a door 1.5m (5ft) wide. Bedding can be sand, sawdust or chopped straw.

Young rheas. There is no need to have a permit to keep them, as there is with emus and ostriches.

The breeding season is from spring to autumn, depending on the weather. A female lays an average of thirty eggs, although if they are removed, the number is nearer forty to fifty. Eggs collected for incubation need to be very clean. Optimum conditions for hatching the eggs are a temperature of 36°C (97°F), and relative humidity of 40 per cent for days 1–32. From day 33–36 the temperature is decreased to 35.5°C (96°F), while humidity is increased to 75 per cent. Once hatched, chicks need a brooding area with a heat lamp. They can feed on ratite chick crumbs with a limited amount of fine grit, going over to pellets when they are older.

Ostriches

Like emus and rheas, ostriches are ratites, and they are the largest birds in the world: a fully grown bird is 2.4m (8ft) tall, and weighs 160kg (350lb). It needs extensive pastures where it can run about freely, and fencing that is 1.8m (6ft) high with five strands of high tensile wire. If members of the public have access to the area, a second line of fencing needs to be erected, along with a warning notice. A permit is also required because ostriches are classified as potentially dangerous. They need a building with sufficient headroom to provide adequate ventilation, and sand or chopped straw for bedding and nesting.

Ostriches are farmed for meat, leather and feathers. In Britain and Europe they became popular in the mid-1990s, but then fell out of favour after breeders and investors were unable to get a worthwhile return on their investments. Concerns were also expressed about welfare standards when it came to managing, transporting and slaughtering the large birds. For the smallholder, farming the ostrich is probably not a feasible option in view of its size, potential dangers and the need for such large, sheltered pastures.

Earthworms

Vermiculture is the farming of earthworms. It is one of those activities that can be carried out on any scale, and has grown in popularity to encompass small domestic, compost units for kitchen waste, as well as farm-scale enterprises for the production of worm compost and bait worms.

The earthworm is a remarkable creature, able to convert organic waste such as decaying vegetation into nutrient-rich compost, while at the same time improving the drainage of the soil in which it is working. The resulting compost is a fine, fertile medium for subsequent plant growth. It is a rich

Ostriches need high, strong fences and plenty of running space.

source of minerals, including nitrates, phosphorus, potassium, magnesium and calcium, making a natural substitute for chemical fertilizers.

The principle of worm composting is to provide a suitable environment (a wormery) in which the worms are provided with bedding in which to breed, and regular supplies of organic waste on which to feed. From here, the resultant compost castings are regularly removed, as well as young worms or egg capsules for the setting up of new units. The wormery can be indoors or outdoors, depending on conditions, and the scale will depend on the enterprise.

There are hundreds of different species of earthworms in the world. For our purposes, we can differentiate them into two groups: the soil-dwelling ones, such as the well-known common earthworm or lob worm, *Lumbricus terrestris*; and manure worms that are the best choice for a wormery. (They live in the organically rich surface layer of the soil, while common earthworms are burrowers, requiring greater soil depths.) Suppliers of worms will supply either one type or a mixture of worms, depending on what they are breeding. Common earthworms are also available for sale. Manure worms commonly found or offered for sale include the following:

Wild Red, *Lumbricus rubellus*: Maroon-red all over, without particularly distinctive rings. It can be found naturally if you leave some cardboard to rot on the soil surface in the garden.
Large Red (Dendra, Blue Nose, Superworm), *Dendrabaena veneta*: One of the largest varieties, and can be recognized by the blue colour at the head end.
Tiger (Brandling), *Eisenia foetida*: Larger than the 'wild' red worm, this has distinctive yellow and maroon-red rings. *Eisenia andreii* is a close relative.

A Household Wormery
To set up a small wormery, you can either make your own or buy a ready-made unit. There needs to be a relatively large surface area in relation to the depth, for it is at the surface that the worms will feed. If it is a DIY bin unit, drill some drainage holes at the bottom and aeration holes along the top sides. Place the container on some bricks to keep it clear of the ground, and ideally have a container underneath to catch excess liquid. This makes a marvellous plant fertilizer when diluted in the ratio of one part liquid to ten parts water. Some bought units are equipped with drainage chips and have a tap so that the liquid can be drained off easily and cleanly.

In the bin, place some damp bedding: suitable materials are compost, torn up and moistened newspaper (soak, then wring out), small cardboard pieces and wet, decaying leaves. Do not use any grass clippings because they will heat up and possibly kill the worms. The bedding should be loosely placed rather than compacted so there is enough air to breathe. Add the manure worms and cover lightly with material such as hessian, carpet underfelt or damp newspaper sheets. A loose-fitting lid to exclude light is also recommended, as long as aeration is not adversely affected. Again, bought units are more convenient in this respect because they have lids which are more effective at keeping out the little fruit flies, *Drosophila meleangaster*, that can be a nuisance.

Water as necessary to keep the whole thing damp but not soggy: around 75 per cent moisture is the amount to aim for. Worms breathe through the skin and will die if it becomes too dry. They will also

A household wormery in a conservatory. It has just had an addition of vegetable parings and teabags.

drown if it is too wet. Bury scraps lightly at the surface to avoid smells, but the worms will gradually pull them down; particularly popular are tea bags, coffee grounds, vegetable and fruit. Avoid meat and fish scraps because of the smell and the risk of attracting house flies. Do not add cat or dog litter faeces because of potential health risks to humans. Citrus fruits should either be avoided or used sparingly because they are very acidic. Little and often is better than adding large amounts of waste at a time, and dense material such as cabbage stalks should ideally be shredded.

Sprinkle on a little garden lime occasionally, not only to avoid smells but also to stop the compost becoming too acidic. Calcified seaweed also works well. Worms operate best at pH 7.0. Dried, crushed eggshells are also a useful source of calcium: they reduce acidity, but provide some of the roughage that the worms need to digest their food.

As a general guideline, a wormery containing 1,000 worms (weighing around 450g (1lb)) can take around 1.4kg (3½lb) of kitchen waste. Start with a small amount at first, and gradually increase it. As the worms breed, the population of the wormery should double in two to three months. It is important to remember that there needs to be sufficient bedding for them to breed, in relation to the amount of food waste. They feed on the latter, but they breed in the former.

Keep the unit in a temperature range of 10–25°C (50–78°F). In winter, it may need to be placed in a more protected place and to have insulation. Avoid direct sun so that it does not over-heat. Periodically remove the top layer of worms and bedding and either start another bin, or replace when the underlying new compost castings have been removed. With a bought unit, follow the manufacturer's instructions, because they will be appropriate to their units.

The harvest is in two forms: the worm castings compost and a new crop of worms. Taking off the top layer where food matter is supplied will bring most of the worms with it. If they are still intermixed, dump the layer on a thick sheet of plastic in the light and place damp newspaper on one side. The worms will migrate upwards and sideways to hide under the newspaper, making removal easy. The underlying compost can then be removed for use in the garden or as seed/potting compost, while the worm layer can be used to start a new wormery.

Worm Farming

On a larger scale, it is more appropriate to construct your own worm beds. The arrangement can be set up in an outbuilding, barn or polytunnel, as long as temperature and shade can be adequately controlled. Outside worm beds or pits are also an option, and one idea is to have breeding worms inside, while 'growers' which are possibly being produced for the fishing bait industry are outside. Other commodities that can be offered for sale are bags of worm compost, breeding worms and zoo supplies. As with any farming venture, it is essential to have researched and obtained assured markets before going into extensive production. Trial it on a small scale first in order to establish strengths and weaknesses.

Edible Snails

Edible varieties of snails are definitely for the speciality market, but a great deal of research is required to ensure that the market is big enough for the degree of work entailed. The Great African Land Snail, *Achatina achatina*, is more productive than other varieties, but is usually associated with intensive production requiring specialized buildings and heating. Escargot varieties are slower growing, and represent the quality end of the market. They are the ones reared in France and include Le Petit Gris, *Helix aspersa major*, and Le Bourgogne (Roman snail), *H. pomatia*. They are reared in more extensive conditions such as polytunnels, although cold weather protection is required. Snails are hermaphrodite, and after two have mated, both lay eggs.

There are many other enterprises that may be worth considering, including aquaculture for fish or crayfish production, cattery or dog-minding services, equine services, and so on. There is an extraordinary range of pursuits to explore. The modern smallholder, untrammelled by old attitudes, is better suited than anyone to make a success of them.

Appendix I: Working in the Countryside

Small farmers are better suited to relationship marketing than are larger farms, whose sheer size requires selling food in an impersonal marketing system. (*Small Farm Digest*, US Department of Agriculture, 2001)

A rural enterprise is quite simply any commercial activity that is taking place in the countryside. It may, or may not, be connected with agriculture, depending upon the individual situation. The activity may be part-time for one or two people, or full-time for one partner and part-time for the other. There are many other possibilities, but behind any activity is the need for somewhere quiet from which to organize it, even if it is just for 'keeping the books'.

In Britain, if it is intended to use an existing building for a new, commercial activity, it may be necessary to obtain 'change of use' permission from the local authority. If a completely new building is required for a business, the local authority must again be approached, this time for actual planning permission, rather than just change of use permission. This is sometimes more difficult to achieve, particularly if an area is designated as a conservation area. Only individual local authorities are in a position to give specific advice on this. If permission is given for the erection of a new building, it is still necessary to acquire 'building permission': this is to ensure that minimum building safety regulations are observed.

Once local authority permission is obtained, a commercial rate will be levied on the premises. It is worth remembering that, although costs of lighting, heating, equipping and maintaining business premises can be offset against tax, it may be necessary to pay capital gains tax in the event of the site being sold. An accountant will advise on this question, as well as on any other problem that may arise with a home business. The USA, with its much larger land space, has fewer restrictions, but the situation varies between different states. Information is available from state, county and city agencies, as well as federal or national ones. They are listed in the telephone directory.

The Home Office

A converted outbuilding, spare room or even a shed can be used as an office, as long as it provides warmth, light, comfort and access to electricity and telephone lines. There are companies who manufacture garden buildings for use in this way, while others specialize in converting existing areas.

A working area needs to be big enough to allow for a variety of activities such as reading, writing, keying in, telephoning, as well as relaxing and thinking. Environment is important! Walls that are painted in light-enhancing and pale matt colours provide soft, clean lines without being soporific. Pictures and photographs on the walls add to the congeniality, while plants, in my view, are essential in any room.

A big noticeboard on which to stick reminders and post-it notes is helpful, while a calendar with large numbers is also useful. Natural light from a window is essential, although it can be a nuisance when it falls on a computer monitor. A blind with vertical slats is more effective than an up-down blind or a curtain because it diffuses the light rather than excludes it.

Essential equipment includes a desk, chair, table, computer with printer and modem, telephone, fax, filing cabinet and waste-paper basket. To these I would add a bookcase, armchair with footrest and coffee table, not to mention tea- and coffee-making facilities. A plain paper copier may not be essential, particularly if you have a low-cost copying facility locally, but it is extremely useful if you need to produce copies in a hurry, or in bulk. Small, personal copiers are now available, or a fax can be used to produce individual copies. A scanner with optical character recognition (OCR) software is most useful for scanning in documents and helping to control the flow of paperwork.

Perhaps most important of all is to have an Internet connection so that e-mails can be sent all over the world for the cost of a local telephone call. A service provider for business users is more expensive than one that caters purely for home users, but it can usually be relied upon to provide a connection at all times, as well as a regularly managed help-line. The world-wide web is an amazing source of information and contacts, as long as the search engines are used to select only those sites that are relevant. (It pays to have up-to-date anti-virus software, as well as that which filters out unsolicited communications or dubious websites.) Finally, having one's own website is important in providing a shop window – but buying and selling by credit card over the Internet is only feasible if secure server facilities are available.

The Farm Shop

It may be necessary to have planning permission from the local authority to open a farm shop, unless an existing building is used. It also depends on whether you are selling your own produce or that from another source. In the case of the latter, permission may be needed and the local authority will advise. In the USA, roadside stands are much more common than they are in Britain, where local authorities generally do not allow direct trading on the side of the road. There are still regulations to be met in the USA, however, and it is worth checking on local, county and state regulations before going to the expense of building a stand.

Farm-gate sales are popular with consumers, but do rely on car access. Adequate car parking and turning space is needed, as well as clear road signs. Good signs make a lot of difference: motorists are far more likely to stop if they see an attractive sign, rather than a quickly scrawled offering. Signs further down the road, warning motorists that the site is coming up, are also necessary. These should be placed in such a way that they are clearly seen from the road, and in plenty of time for the driver to react safely, and not with a last-minute swerve. Equally important are opening times, making it clear, from the road, whether it is open or closed. There is nothing more annoying than turning in through someone's gate in response to a sign outside and then finding that the site is closed.

If selling produce by weight, be aware of 'weights and measures' legislation, as well as labelling and trade descriptions. Food products are also covered by food safety regulations. Contact the local Environmental Health Officer for further information.

Clear notices that can be seen from the road are essential for farm shops.

Local authority permission may be required for erecting signs on the roadside, and officials will also wish to satisfy themselves that there is no traffic hazard involved.

Pick your Own

The 'pick your own' enterprise is popular with growers and public alike (in the USA it is called 'U-pick'). The strawberry crop is the most common, with family parties often going to pick at weekends, then taking their baskets to be weighed at the check-out point of the farm shop. The experience of many growers is that it is better to provide standard containers and to sell by weight rather than volume. This is also clearer from the point of view of the customer. There are customers who prefer not to

Pick-your-own strawberries, a popular weekend activity in the summer.

pick their own so it is important to have containers that are full and ready for sale in the shop.

Again, a car park with adequate turning space is required, as well as road signs warning that the site is just ahead. Signs on the site itself should make it clear where the public walkways are, and these may need to be roped off so that the pickers are in no doubt as to the direction to follow.

Public liability insurance is essential, but it is also important to make clear to parents that they are responsible for their children while on site. Dogs should not be allowed in at all and a clear sign in the car park should indicate that they are to be left in the cars. In hot weather, some visitors may need to be reminded that the car windows should be left open or the animals may suffer heatstroke.

Crops suitable for picking by the public are generally fruit such as strawberries, raspberries and plums, and vegetables such as runner beans (pole beans), dwarf or French beans, and sweetcorn.

The Farm Park

Some smallholders, particularly those who keep a range of traditional breeds of livestock and poultry, have successfully opened small farm parks to the public. Having a range of rural crafts such as spinning and weaving on view is also popular from the educational point of view. Toilet and refreshment facilities are essential, while a small gift shop is popular. Talking to local schools and organizations is often useful so that specific tours can be arranged. Local publicity is vital: letting the local media know what is happening ensures that people know of your existence, and local advertising can then follow.

Again, public liability insurance is essential, as

well as good signposting and a car park. It is important to remember that if the public, and particularly children, have close access to animals, a notice should be erected advising them to wash their hands after touching them.

Bed and Breakfast

Providing bed and breakfast accommodation for tourists is a popular way of generating an income in the countryside, but it is necessary to work out how much time and resources are required. This will affect the scale of the business, and whether the service is available all year round or just during the holiday season. Planning permission may be needed for a small bed and breakfast business, but it depends on individual circumstances. If it is needed, then factors such as car parking facilities, effects on neighbours, and the number of bedrooms let are considered. Further information is available from the local council. Help and information, as well as publicity, is available from local and regional tourist boards.

If more than six people, including guests, staff and children, are accommodated, a fire certificate from the local fire authority is required. Even if a certificate is not necessary, it is still a good idea to discuss plans with the fire authority. All electrical systems must be maintained to avoid danger to those who use the premises, including guests. If accommodation is offered for up to six people, business rates do not apply, provided that the B&B is no more than a subsidiary use of the home. This is determined by the local valuation officer of the tax authority. Further details are available in a free leaflet entitled 'Bed and Breakfast and the Business Rates' from the regional tourist board.

All livestock respond to attention, but it is important to wash the hands after touching them. In farm parks a notice to this effect is required.

If meals are served, you will need to register with the local authority and observe food safety law. All guests over sixteen years old must be registered, with details kept for at least twelve months. For laws and regulations regarding full- or part-time staff, contact the local branch of the Department of Employment.

Supplying customers direct at a local farmer's market.

Organic Production

This is a sector that the small producer is ideally placed to serve. Fruit, vegetables, eggs and meat can all be sold as organic, if they meet the standards, and there is an increasing consumer demand for fresh produce without chemical residues. There is a common organic standard throughout the European Union, but that of the USA is still developing, and generally speaking is far less stringent than that of Europe. Details are available from USDA or the Organic Farmers' Marketing Association.

In Britain, details can be obtained from UKROFS. It is necessary to be registered with one of the certification bodies, such as the Soil Association or Organic Farmers and Growers so that the site can be inspected. The standards cover growing methods, housing and management. No genetically modified materials or in-feed medications are allowed in livestock feeds. There is also a conversion period while land is changing over to organic status.

If only an occasional surplus is sold, it may not be possible to justify the cost of organic registration, even though the produce is organic in all but name. Here, concentrating on local sales is the best option, at the farm gate or at a local farmer's market.

The Farmer's Market

These are ideal places at which to sell produce to local customers, and are popular in Europe and the USA. They are normally held at specific venues and at regular times, such as the first and third Saturday of the month, or every week. To sell in this way requires good organization to ensure that there are no seasonal dips in supplies. Time and resources are also required for transporting the produce to the market, as well as manning the stall.

Rare Breeds Marketing

In Britain, the Rare Breeds Survival Trust has set up a marketing scheme to provide breeders with sales' outlets. The idea behind it is to encourage the breeding of traditional and endangered breeds, and it has proved popular with butchers and customers alike.

Appendix II: Regulations

The regulations that are in force in relation to livestock are essentially to do with registration and identification, health, transport, movement and records, as well as slaughtering and selling. In Britain the local Animal Health Office (AHO) and local branch of the Department of the Environment, Food and Rural Affairs (DEFRA) are the relevant organizations to contact, while in the USA it is the local extension service of the United States Department of Agriculture (USDA). The requirements obviously vary to some degree: those below apply to Britain.

Registration: Anyone with livestock such as pigs, sheep, cattle and goats must register with the AHO of the local Trading Standards office and obtain advisory information at the local DEFRA office (see local telephone directory).

Identification: Livestock must be identified with a tattoo or ear tag, and appropriate records need to be kept of each animal. For cattle there is also an additional passport that accompanies each animal.

Welfare: There are Codes of Recommendations for the Welfare of Livestock for each type of animal. They are available from DEFRA or AHO offices.

Movement of livestock: Livestock moved from one site to another must be accompanied by a movement licence. This is called the Schedule 1 Holding Movement Record, and includes the name and address of the person keeping the record, the number and identification marks of the animals, the date of the movement, and the premises the animals are to be moved from, and those they are to move to (*see* page 178). It is possible to issue your own licences: blank copies are available either from the AHO, or by getting a book of tear-out licences from the local NFU office. A blank copy can be photocopied for producing your own.

Transport of livestock: The Animal Health Act and the Welfare of Animals During Transport specify the welfare conditions and requirements for transporting, loading and unloading animals. Trailers must be suitable for the number and type of livestock, and provide safe, non-slip access. The trailers must be cleaned and disinfected after use.

Slaughtering of livestock: Animals must be slaughtered at a licensed abattoir by a licensed slaughterer. Meat is inspected, butchered and packed according to the Fresh Meat Hygiene and Inspection Regulations. Casualty animals that have to be killed on the farm must be killed humanely and quickly by an experienced person. The vet or AHO will provide advice and help.

Medical records: It is necessary to have a Schedule 2 Veterinary Medicine Administration Record. This is a record of veterinary products administered to all farm livestock. Furthermore, records must be made within seventy-two hours of the administration of a medicinal product to an animal. A veterinary product is a product that has been licensed for veterinary use and has a product licence number. There are two categories: POM, available on prescription only from a vet; and PML that can be bought from a vendor who is licensed to sell it. It is illegal for anyone who is unregistered to sell veterinary products.

Specified withdrawal periods are laid down between the end of treatment and the slaughter of animals or the taking of eggs or milk for human consumption. If animals are sold before the end of the withdrawal period, the purchaser or auctioneer must be informed. In commercial enterprises, records should be kept for a minimum of two years after treatment. The details to be recorded are as follows: date purchased, name, quantity and supplier of medicine; identity and number of animals treated; amount given; date treatment finished and date when withdrawal period ended; name of person administering the medicine. The records can be inspected by DEFRA officers.

Notifiable diseases: Some diseases such as foot and mouth disease, swine fever, sheep scab and fowl pest must be notified to DEFRA. A full list is available from the Animal Health Division (A), Hook Rise South, Tolworth, Surbiton, Surrey KT6 7NF.

Zoonoses: Some diseases, such as salmonella, campylobacter and cryptosporidium, can be transmitted from livestock to humans. There is a complete list in the Health and Safety publication entitled *The Occupational Zoonoses*; this is available from DEFRA.

Dangerous wild animals: In order to keep ostriches, emus and wild boar it is necessary to have a permit from the AHO and to display a warning notice to the public if they have access to the site.

Selling produce: Meat that has been prepared according to the Fresh Meat Hygiene and Inspection Regulations can be bagged and returned to the owner for use or sale. To sell the meat, the producer must be registered with the local Environmental Health Department who will need to inspect the premises to ascertain that Food Safety Act requirements are being met, as well as the labelling and description requirements of the Trades Description Act.

Where eggs are being sold, there is no need to register as long as they are sold ungraded to callers, friends and neighbours. If they are to be graded and packed, according to the sizes shown on page 131, the premises must be registered and inspected by the Regional Egg Marketing Inspector who will provide the necessary information. These details are also on the DEFRA website, at www.defra.gov.uk

Where cow's milk is sold it is necessary to purchase a 'milk quota' and be a registered producer. Details are available from DEFRA. The sale of milk and dairy products are covered by the Milk and Dairy Regulations, the Food Safety Act, Weights and Measures and Trades Description regulations. The premises must be registered with the Environmental Health Department who will inspect the milking, dairying and packaging areas. Milk hygiene, testing and treatment procedures must be complied with, as well as other legislation covering as packaging, labelling and descriptions. Further information is available from DEFRA and the local Environmental Health Office.

Appendix III: Coping with Pests the Organic Way

Avoiding pests and problems is always the best policy, with attention given to rotation of pasture or growing area, as well as to animal welfare, hygiene and veterinary protection. Pyrethrum is derived from the *Chrysanthemum coccineum* plant. It is effective against aphids and insect pests, but it also kills beneficial insects, so care should be taken with its application. The following is by no means a

Problem	Remedy
Aphids	Encourage natural predators such as ladybirds (ladybugs), lacewings and hoverflies by planting marigolds, nasturtiums and fennel. Pyrethrum is also effective.
Blackfly	Often found on growing tips of broad beans, blackfly aphids can be discouraged by pinching out tops of plants. Spraying with a liquid seaweed preparation acts as a deterrent as well as foliar feeding the plant. Pyrethrum is effective.
Botrytis (grey mould)	Found as brown spots with grey mould on seedlings and leaves, this thrives in cold, damp conditions. Remove affected plants, reduce watering and improve ventilation.
Apple sawfly	Sawfly larvae feed near surface of apple causing raised scars. Remove affected fruit as soon as seen. Spray tree with Pyrethrum as soon as the blossoms fall.
Asparagus beetle	Greenish grubs, black and yellow beetles. Remove when seen. Use Pyrethrum.
Cabbage root fly	Sow your own plants. Lime soil adequately (pH 7.5). Protect plants with collars.
Carrot fly	Erect a 60cm (2ft) high barrier around the carrot bed to stop the low-flying flies.
Caterpillars	Cabbage white caterpillars on brassica. Pick them off. Pyrethrum can be used. The biological control *Bacillus thuringiensis*, is a bacterium that stops them eating.
Clubroot	Avoid bought-in brassica plants. Sow your own. Add lime to soil (pH 7.5). Use collars.
Codling moth	Moths lay eggs in developing apples. First sign is a maggot in the apple. Hang up one pheromone trap for every four trees.
Flea beetles	Beetles often found on turnip plants. Put a layer of oil or grease on a piece of wood and pass it over the tops of the leaves. The beetles jump up and stick to the grease.
Fungus leaf spot	Follow crop rotation. Remove affected areas. Spray with dispersible sulphur.
Onion fly	Grow onions from sets rather than from seed. Plant near carrots.
Red spider mite	Plant marigolds in greenhouse beds. Introduce biological control *Phytoseiulus persimilis*, a predatory mite that controls the red mite.
Sawfly	Sawfly caterpillars decimate leaves, especially on gooseberry shrubs. Encourage robins, for they eat them. Pyrethrum is also effective.
Slugs	Let ducks clear the area in winter. Encourage frogs and toads that eat slugs by having a pond. Set up some 'beer traps'. Use a non-poisonous product such as Ferosan. Coffee solution is also effective.
Whitefly	A pest in greenhouses. Hang up sticky yellow traps. Biological control, *Encarsia formosa* is a parasitic wasp that controls them.
Woolly aphids	Aphid colonies that cover themselves with a white, protective (woolly) coating on twigs and stems. Scrape off or paint with methylated spirit. Pyrethrum is also effective.

comprehensive list, but it does offer some suggestions that may be useful to those who are concerned with more natural systems.

As far as controlling pests and problems where animals are concerned, it is important to remember that while organic standards prohibit the indiscriminate use of in-feed antibiotics, their use when real need demands it, is of course allowed. Within the EU, antibiotics can only be prescribed by a vet.

Coccidiostats are allowed in poultry starter rations. If there is a subsequent problem with coccidiosis in an older bird, Amprolium can be used.

The control of internal worms is encouraged to be by effective pasture rotation and periods of fallow so that their life cycle is broken. Routine protection is allowed after consultation with the certification body. All ruminants can be treated with Morantel citrate (Exhelm for sheep and Paratect Flex or Panacur for cattle).

Where vaccines need to be used, then multi-vaccines (two-in-one or four-in-one) are preferred. The vaccine choice and use should only be used in consultation with the vet.

For foot rot conditions, the traditional remedies of copper sulphate solution, zinc sulphate solution and iodine are effective.

Outdoor calving and ensuring that calves receive colostrum within six hours of birth is recommended as a way of avoiding scouring. Where it occurs, oral re-hydration with a glucose electrolyte solution is recommended. Other veterinary products may be given on the recommendation of the vet. Where there is a problem of husk on a smallholding (a form of bronchitis caused by lungworms) the oral husk vaccine Dictol can be given to calves before they go out on pasture.

Every effort should be made to avoid mastitis, with practices such as teat wiping, fore-milking and regular milk and machine testing. Collodion can be used to seal and protect the teats during drying off. Iodophore is effective for teat lesions and disinfection. Uddermint is useful in the early stages, as well as a homoeopathic nosode in the drinking water. In severe cases, an antibiotic preparation from the vet should be used.

In organic systems, the withdrawal periods after medications have ceased are generally longer than those defined by the product licence, and are usually at least two weeks.

Glossary

Abattoir — Slaughterhouse for livestock.

Ad lib feeding — Feeding at any time.

Animal passport — Required documentation for identifying cattle in Britain.

Annual — Plant that grows, flowers and sets seed in one season.

Aspergillosis — Fungal infection of the lungs.

Antibodies — Protective agents in the bloodstream.

Ark — Movable shelter for livestock on pasture.

Ash — Mineral content of proprietary feeds.

Bantam — Naturally small poultry.

Biennial — Plant that grows, flowers and sets seed over two seasons.

Blanching — Excluding light from plants.

Breeder — Male or female livestock or poultry kept for breeding.

Broiler — Table chicken.

Broken mouthed — An older sheep with missing teeth.

Brooding — Providing protected conditions for young birds after hatching.

Buck — Male goat or male rabbit.

Bullock — Young male or castrated bull.

Calving — Giving birth to calves.

Colostrum — First milk rich in nutrients and antibodies produced after birth.

Coppicing — Partial cutting of a tree to produce new shoots.

Dairy ration — A proprietary feed formulated for the needs of milk-producing animals.

Disbudding — Practice of cauterizing horn buds so that horns do not grow.

Drenching — Administering medication in liquid form.

Docking — Shortening the tail, e.g. of lambs.

Dutch barn — Barn without sides, used for hay and straw storage.

Farrowing — Giving birth to piglets.

Flushing — Bringing up to peak condition for mating.

Fodder — Fibrous food for livestock.

Fore-milk — First milk extracted from the teats during milking, and discarded.

Free range — Allowing unrestricted access to outside pasture during daylight hours.

Gestation — Period of pregnancy.

Gilt — Young, unmated female pig.

Harrowing — Breaking down into a fine tilth soil that has already been ploughed.

Heifer — Young, unmated cow.

Hybrid — Livestock or poultry developed from several breeds or strains.

Keet — Young guinea fowl.

Kibbling — Chopping grains to a smaller size.

Kidding — Giving birth to goat kids.

Kindling — Giving birth to rabbits.

Lactation — Period of milk production following birth.

Lambing — Giving birth to lambs.

Litter — Material such as chopped straw used as bedding or floor covering in livestock and poultry housing. Also a group of newly born animals.

Maintenance ration — Feed ration given to cater

	for basic requirements.
Mouse guard	Obstacle to prevent mice getting into beehives.
Mulch	Layer of material such as compost to protect underlying soil.
Notifiable disease	One that is required by law to be notified to the authorities.
Perennial	Plant that lives for many years, flowering every season.
pH value	Potential hydrogen level, indicating the degree of acidity or alkalinity.
Pick your own (U-pick)	Allowing members of the public access to a crop so that they can pick and buy it direct.
Pole barn	Type of field shelter.
Polled	Hornless breed.
Pop-hole	Exit or entrance hole to a poultry house.
Poult	Young turkey.
Production ration	Supplementary feeding on top of basic ration, to cater for milk or egg production.
Pullet	Young female bird that has not yet begun laying.
Pure breed	One that always produces young like itself when crossed with its own breed.
Queen excluder	Obstacle board to stop the queen bee going up into the area where honey is produced.
Ringing	Identifying by means of a leg ring, usually poultry or rabbits.
Roll bar	Safety bar on a tractor in

	the event of turning over.
Scouring	Diarrhoea, particularly in young animals.
Sire	Father or male line.
Starter	A dairy culture for producing yoghurt or cheese.
Store animals	Animals kept for breeding or future fattening.
Tagging	Fitting an animal with an identifying tag, often on the ear.
Tattooing	Alternative method of identifying a specific animal.
Terminal sire	Male for crossing with other breeds for utility purposes, e.g. meat production.
Trap nest	One that allows access but not exit until eggs can be identified with individual birds.
Tupping	Mating of sheep with ram.
Utility breed	One kept for production rather than for showing.
Vermifuge	Medication for expelling parasitic worms.
Weaning	Process of curtailing an all-milk diet in a young animal.
Worming	Taking action to expel parasitic worms.
Yarding	Practice of housing livestock, particularly cattle, in a straw yard with shelter in winter.
Zoonoses	Animal disease-causing organisms that can also affect man.

Sources

Publications

Broad Leys Publishing Ltd. www.kdthear.btinternet.co.uk.
Case, Andy *Starting with Pigs* (Broad Leys Publishing Ltd, 2001).
Castell, Mary *Starting with Sheep* (Broad Leys Publishing Ltd, 2001).
Country Smallholding magazine. CML. Publications. www.countrysmallholding.com
Thear, Katie *A Kind of Living* (London, 1983).
Thear, Katie *Part-Time Farming* (Ward Lock, London 1982).
Thear, Katie *Starting with Chickens* (Broad Leys Publishing Ltd, 1999).
Thear, Katie *The Family Smallholding* (Batsford, London 1983).

Advisory Organizations

DEFRA (formerly MAFF). Tel: (Helpline) 08459 335577. Publications: 08459 556000
ADAS. Tel: 08457 766085
FAWC (Farm Animal Welfare Council). Tel: 020 7904 6531
Freedom Food Ltd. RSPCA, The Manor House, Causeway, Horsham, West Sussex RH12 1HG. Tel: 01403 223154

Societies

British Alpaca Society. Tel: 01403 786814
British Angora Goat Society. Tel: 01789 841219
British Beekeepers' Association. Tel: 02476 696679
British Commercial Rabbit Association. Tel: 01270 780248
British Coloured Sheep Breeders' Association. Tel: 01873 890212. www.bcsba.org.uk
British Waterfowl Association. Tel: 01564 741821. www.waterfowl.org.uk
Centre for Alternative Technology. Tel: 01654 702400. www.cat.org.uk
Henry Doubleday Research Association (HDRA). Tel: 024 7630 3517
Humane Slaughter Association. Tel: 01582 831919
National Sheep Association. Tel: 01684 892661
NFU National Farmers Union. Tel: 01572 824686
Organic Farmers and Growers Ltd. Tel: 01353 722398
Pygmy Goat Club. Tel: 01822 834474
Rare Breeds Survival Trust. Tel: 024 7669 6551
Rhea and Emu Association. Tel: 01455 823344
The British Goat Society. Tel: 01626 833168
The British Houserabbit Association. www.houserabbit.co.uk
The British Llama and Alpaca Association. Tel: 01372 458350. www.llama.co.uk and www.alpaca.co.uk
The British Pig Association. Tel: 01923 695295
The British Rabbit Council. Tel: 01636 676042
The Poultry Club of Great Britain. Tel: 01205 724081
The Soil Association. Tel: 0117 929 0661
Timber Grower's Association. Tel: 01981 240250. www.timber-growers.co.uk
UKROFS (United Kingdom Register of Organic Farm Standards). Tel: 020 7238 5915

Suppliers

Home

Clearview Stoves. Tel: 01588 650401. Multifuel stoves.
French, Flint and Ormco Ltd. Tel: 020 7403 1733. Glass and plastic storage containers.
Natural Casing Company Ltd. Tel: 01252 850454. Sausage-making supplies.
Sutcliffe Electronics. Tel: 01233 634191. Intruder alarms.
The Ecology Building Society. Tel: 0845 674 5566. www.ecology.co.uk.
The Handweavers Studio and Gallery Ltd. Tel: 020 8521 2281. www.handweaversstudio.co.uk. Fibrecraft supplies.
Vigo Ltd. Tel: 01823 680844. Fruit presses and crushers.

Land

A.P. Equipment. Tel: 01233 712231. Smallholding equipment

Agralan. Tel: 01285 860015. Organic controls
Ashridge Trees. Tel: 01837 89099. Hedging and trees.
Buckingham Nurseries. Tel: 01280 813556. www.bucknur.com. Hedging and trees.
Bunce (Ashbury) Ltd. Tel: 01793 710212. Garden machinery.
CJ Industries. Tel: 01239 615300. www.cjindustries.co.uk. Garden machinery.
Chipperfield Garden Machinery. Tel: 01923 269377. Garden machinery.
Citadel Products. Tel: 01789 297456. Polytunnels.
Deacons Nursery. Tel: 01983 522243. Fruit trees.
Ferryman Polytunnels. Tel: 01363 83444. Polytunnels.
Gordon Hill. Tel: 01404 812229. Fast-growing poplars and willows.
Green and Carter. Tel: 01823 672365/672950. Ram pumps.
Jinma (UK). Tel: 01952 618182. www.jinma.co.uk. Small tractors.
Knowle Nets. Tel: 01308 424342. Fruit cages.
Newlandowner Management Services Ltd. Tel: 01283 585410 www.newlandowner.co.uk. Advice for smallholders.
Nutriculture Ltd. Tel: 01704 822586 Fax: 01704 822678. Hydroponics equipment.
Nutwood Nurseries. Tel: 01326 564731/573593. Nut trees.
Stratford Power Garden Machinery. Tel: 01789 294839. Garden machinery.
The Willow Bank. Tel: 01594 861782. www.telecentres.com/Willow_Bank. Fast-growing willows.
Tracmaster Ltd. Tel: 01444 247689. Two-wheeled tractors.
Walcot Nursery. Tel: 01386 553697. Organic fruit trees.
Wychwood Tunnels. Tel: 01452 790650. Polytunnels.

Poultry and Livestock
Aliwal Incubators. Tel: 01508 489328.
Allen and Page. Tel: 01362 822902. www.allenandpage.com. Free range and organic feed suppliers.
Amtex. Tel: 01568 610900. Smallholding equipment.
Ascott Smallholding Supplies. Tel: 01691 690750. www.ascott.org.uk
Atlantic Superstore. Tel: 01986 894745. www.atlanticcountrysuperstore.co.uk. Smallholding equipment.
Beefi Startin and Roxan ID. Tel: 01750 22940/30. Poultry and livestock identification systems.
Brinsea Products Ltd. Tel: 01934 823039. www.brinsea.co.uk. Incubators.
Chick Equip. Tel: 01476 585259. Poultry equipment.
Cliverton Insurance Brokers. Tel: 01263 860388. Insurance for smallholdings.
Cyril Bason (Stokesay) Ltd. Tel: 01588 673204/673242. Poultry and poultry equipment.
D.M. Harrison Homoeopathic Pharmacy. Tel: 01974 241376. Homoeopathic medicines for livestock.
Danro Ltd. Tel: 01455 847061/2. Labels for farm products.
Electranets Ltd. Tel: 01452 617841/864230. Electric netting for poultry and livestock.
Electric Fencing Direct. Tel: 01732 833976. www.electricfencing.co.uk
Forsham Cottage Arks. Tel: 01233 820229. www.forshamcottagearks.co.uk. Poultry housing.
G.A. and M.J. Strange. Tel: 01225 891236. Electric netting for poultry and livestock.
Galen Homoeopathics. Tel: 01305 263996.
Gardencraft. Tel: 01766 513036. Poultry housing.
Greenlands Insurance. Tel: 01726 843597. www.greenlands.co.uk. Insurance for smallholders.
Hengrave Feeders Ltd. Tel: 01284 704803.
Interhatch. Tel: 0700 4628288. Incubators and poultry supplies.
Intervet. Tel: 01223 420221. Poultry vaccines.
Lindasgrove Arks. Tel: 01283 22990. Poultry housing.
Lifestyles UK Ltd. Tel: 01527 880078. Poultry housing.
Littleacre Products. Tel: 0121 308 2251. Poultry housing.
Meadows' Poultry Supplies. Tel: 01733 380288. Poultry housing and equipment.
M.R. Harness. Tel: 01299 896827. Halters for livestock.
Ottery Insurance Services. Tel: 01404 81349. www.ottery.co.uk
Oxmoor Smallholder Supplies. Tel: 01757 288186.
Parkland Products. Tel: 01233 758650. www.parklandproducts.co.uk. Poultry autofeeders.
Peter J. Collin. Tel: 01638 750665. www.frenchall-goats.co.uk. Equipment for goatkeepers and smallholders.
Pintail Sporting Services. Tel: 01794 524472. Poultry equipment.
RENCO. Tel: 01453 752154. www.renco-netting.co.uk. Electric netting for poultry and livestock.
Small Acres Supplies. Tel: 01938 820495. Smallholding equipment.
Smiths Sectional Buildings. Tel: 0115 925 4722.
Solway Feeders Ltd. Tel: 01557 500253. www.solwayfeeders.com. Poultry equipment.
Southern Aviaries. Tel: 01825 830930. Aviary and poultry equipment.
Stock Nutrition. Tel: 01362 851200. Poultry plucker.
The Domestic Fowl Trust. Tel: 01386 833083. www.mywebpage.net/domestic-fowl-trust
The Natural Fibre Company. Tel: 01570 422956. Contract spinning service.
The VPP Co. Ltd. Tel: 01706 358626. Plastic poultry housing.
W. & H. Marriage and Sons Ltd. Tel: 01245 354455. Poultry and livestock free range and organic feeds.
Woodhurst Gardenfowl. Tel: 01487 822356. Poultry housing and equipment.
Woodside Poultry and Livestock. Tel: 01582 841044. Poultry centre, housing and equipment.

USA sources
Organizations
US Department of Agriculture (USDA). www.usda.gov
USDA Agricultural Marketing (Farmers' Markets). Tel: 1-800-384-8704
Agricultural Network Information. www.agnic.org
American Bantam Association. Tel: 973-383-6944
American Boer Goat Association. Tel: 1-800-414-0202
American Cheese Society. Tel: 415-661-3844
American Dairy Goat Association. www.adga.org
American Dexter Cattle Association. www.dextercattle.org
American Farm Bureau Federation. www.fb.org
American Food Safety Standards. www.foodsafety.gov
American Livestock Breeds Conservancy. www.albc-usa.org
American Meat Goat Association. Tel: 919-676-1917
American Poultry Association. Tel: 508-473-8769
American Pastured Poultry Producers' Association. Tel: 715-723-2293
American Rabbit Breeders' Association. Tel: 309-664-7500
American Romney Breeders' Association. www.americanromney.org
Appropriate Technology Transfer for Rural Areas. www.attra.org
Dairy Goat Journal. Tel: 414-593-8385
Icelandic Sheep Breeders of North America. www.isbona.com
Kerr Center for Sustainable Agriculture. www.kerrcenter.com
Missouri Alternatives Center. www.agebb.missouri.edu
Mohair Council of America. Tel: 915-655-3161
National Angora Rabbit Breeders' Association. Tel: 320-762-0376
National Pygmy Goat Association. Tel: 425-334-6506
North American Black Welsh Mountain Sheep Club. www.blackwelsh.com
North American Shetland Sheepbreeders' Association. www.shetland-sheep.org
Northern Nut Growers' Association. Tel: 717-938-6090
Organic Farmers' Marketing Association. www.web.iquest.net
Pennsylvania State University (Agricultural Alternatives). www.agalternatives.cas.psu.edu
Produce Marketing Association. www.aboutproduce.com
Rodale Organic Institute. www.rodaleinstitute.org
Small Farm Digest. www.reeusda.gov/smallfarm
Small Farm Resources. www.farminfo.org
Small Farm Today magazine. www.smallfarmtoday.com
Texel Sheep Breeders' Society. Tel: 815-998-2359
The International Llama Registry. www.lamaregistry.com
University of Nebraska Cooperative Extension. Tel: 402-472-2966
University of North Carolina Extension. Tel: 910-334-7500
Utah State University. Tel: 801-797-1000

Suppliers
Acadiana Aviaries. Tel: 318-828-5957. Waterfowl and quail.
Bend Tarp & Liner Inc. Tel: 800-280-0712. Pond liners.
Brower. www.browerequip.com. Pasture pens and feed equipment.
Common Sense Fence. www.geoteking.com. Fencing
Endurance Net. Tel: 1-800-808-6387. All types of netting.
Heritage Building Systems. Tel: 800-643-5555. Steel farm buildings.
Hilltop Nurseries. Tel: 800-632-2951. Fruit trees.
Hoegger Supply Co. Tel: 800-221-4628. Goat and dairying equipment.
Humidaire Incubator Co. Tel: 513-996-3001. Incubators.
Jordan's Seeds. Tel: 651-738-3422. Bulk vegetable seeds.
Lyon Electric Co. Tel: 760-749-6829. Incubators.
MO River Tire Co. Tel: 660-882-8473. Small tractors.
Murray MacMurray Hatchery. Tel: 1-800-456-3280. Breeds of all fowl, waterfowl and game birds.
New England Cheesemaking Supply Co. Tel: 413-628-3808.
Organic Seed Co. www.organicseed.com
Orscheleln Farm & Home. www.orschelnfarmhome.com. Farm equipment.
Port-A-Hut. www.port-a-hut.com. Portable ark shelters.
Purina Mills. www.purinamills.com. Animal feeds.
Rural Property Bulletin. www.ruralproperty.net
Stachowski Alpacas. www.alpacaworld.com. Alpacas.
Stromberg's Chicks. Tel: 1-800-720-1134. Chickens and game birds.
Sunrise Aviaries. Tel: 510-254-7113. Quail.
Woods-N-Water Inc. Tel: 912-864-7799. Cultivators.

Index

Other farming titles from The Crowood Press

ISBN	Title	Author	Price
1 86126 643 X	*Calf Rearing*	Bill Thickett, Dan Mitchell and Bryan Hallows	£14.99
1 86126 479 8	*Calving the Cow and Care of the Calf*	Eddie Straiton	£18.99
1 86126 383 X	*Cattle Ailments*	Eddie Straiton	£16.99
0 85236 149 1	*Cattle Footcare & Claw Trimming*	E. Touissant Raven	£14.95
1 86126 402 X	*The Domestic Duck*	Chris and Mike Ashton	£19.95
1 86126 271 X	*Domestic Geese*	Chris Ashton	£19.99
1 85223 650 7	*Ducks and Geese*	Tom Bartlett	£9.99
0 85236 350 8	*Forage Conservation and Feeding*	Frank Raymond and Richard Waltham	£15.00
0 85236 235 8	*Goat Farming*	Alan Mowlem	£15.95
1 85223 912 3	*Goats*	Edward Ross	£9.99
0 85236 347 8	*Goats of the World*	Valerie Porter	£24.95
1 86126 174 8	*Honey Bees*	Ron Brown	£9.99
0 85236 543 8	*Improved Grassland Management*	John Frame	£21.95
1 86126 721 5	*Lameness in Sheep*	Agnes Winter	£16.99
1 86126 574 3	*A Manual of Lambing Techniques*	Agnes Winter and Cicely Hill	£14.99
0 85236 188 2	*Modern Shepherd*	Dave Brown and Sam Meadowcroft	£14.00
1 85223 838 0	*Organic Farming & Growing*	Francis Blake	£12.99
1 85223 754 6	*Pigs*	Neville Beynon	£11.99
1 85223 755 4	*Poultry*	Carol Twinch	£9.99
1 86126 261 2	*Poultry Farmer's & Manager's Veterinary Handbook*	Peter W. Laing	£18.99
1 86126 553 0	*Practical Cheesemaking*	Kathy Biss	£9.99
1 86126 389 9	*Practical Goat Keeping*	Alan Mowlem	£16.99
1 86126 388 0	*Practical Pig Keeping*	Paul Smith	£18.99
1 86126 010 5	*Practical Poultry Keeping*	David C. Bland	£14.99
1 86126 049 0	*Practical Beekeeping*	Clive de Bruyn	£24.95
1 86126 163 2	*Practical Sheep Keeping*	Kim Cardell	£16.99
1 85223 835 6	*Profitable Free Range Egg Production*	Mick Dennett	£10.99
0 85236 540 3	*Resource Management – Farm Machinery*	Andrew Landers	£14.95
0 85236 542 X	*Resource Management – Hedges*	Murray Maclean	£9.99
0 85236 559 4	*Resource Management – Soil*	Bryan Davis, David Eagle and Brian Finney	£15.00
1 85223 828 3	*Sheep*	Edward Hart	£9.99
1 86126 397 X	*Sheep Ailments*	Eddie Straiton	£18.99
1 86126 235 3	*Sheepkeeper's Veterinary Handbook*	Agnes Winter and Judith Charnley	£18.99
1 86126 359 7	*Turkeys*	David C. Bland	£10.99

For a fully up-to-date list of our titles, visit www.crowood.com